The BIG PUZZLE BOOK of AREA MAZES

300 Mind-Bending Math Puzzles in Five Challenge Levels

NAOKI INABA and RYOICHI MURAKAMI

THE EXP

NEW YORK

The Experiment, LLC
220 East 23rd Street, Suite 600
New York, NY 10010-4658
theexperimentpublishing.com

THE EXPERIMENT and its colophon are registered trademarks of The Experiment, LLC. Many of the designations used by manufacturers and sellers to distinguish their products are claimed as trademarks. Where those designations appear in this book and The Experiment was aware of a trademark claim, the designations have been capitalized.

The Experiment's books are available at special discounts when purchased in bulk for premiums and sales promotions as well as for fund-raising or educational use. For details, contact us at info@theexperimentpublishing.com.

ISBN 978-1-61519-924-2
Ebook ISBN 978-1-61519-925-9

Cover design by Beth Bugler
Series design by Sarah Schneider
Text design by Jack Dunnington
Translation by Erica Williams | Paper Crane Editions

Manufactured in the United States of America

First printing December 2022
10 9 8 7 6 5 4 3 2 1

BRAIN TRAINING WITH AREA MAZES

As long as I keep my brain active, does it matter how?

What comes to mind when you hear the word "training"? Most likely a "workout" in which you use machines and barbells to strengthen your muscles. But today, more and more of us are "brain training"—stimulating the brain so that its function will not decline. It's a daily step you can take for your health—just like brushing your teeth.

Everything you do uses your brain somehow, so you might think it is enough to just focus your mind on any task. However, some activities are better than others.

Puzzles with math and logic in mind

El Camino is a science and math coaching school for students from first grade through high school. To prepare our elementary school students for the Mathematical Olympiads, we give them puzzles that strengthen their logical and creative thinking.

Saying "let's solve a puzzle!" appeals to the children much more than "let's do a math problem." The students tackle the puzzles enthusiastically—having fun, developing their abilities, and deepening their interest in mathematics all at once.

We invented "area mazes" (*menseki meiro*) especially for our third-grade students. Area mazes can seem impossible to solve without using fractions and decimals. However, our third-grade students haven't learned those techniques yet, so of course you can solve them using only whole numbers! The challenge is to work out *how*. It takes more than calculation: It takes logic, spatial reasoning, and wits.

Getting the answer *without* doing any complicated math is what makes area mazes fun!

Isn't it too late for my brain to grow?

Let me return to the original topic. When our students' parents ask me how to train their own brains, I recommend the puzzles we use at El Camino. You might say, "What? But those are for kids!" Yes, that is true—and that's exactly my point.

The young mind is a flexible "blank slate," whereas the adult brain becomes inflexible. That's why many puzzle books designed for adults fall short: The puzzles tend to follow a pattern. Once your brain gets used to it, you will focus on spotting and applying the pattern—rather than thinking creatively.

To use our metaphor of strength training, doing that kind of puzzle is something like rehab. Repeating an action over and over again will keep your brain active and maintain its function to some extent. However, it cannot rejuvenate your brain: It may put off aging, but it won't make you younger.

Area mazes are different: They are designed to develop flexible thinking—youthful thinking, if you will. They cannot be solved by repeating a process. You often need a stroke of inspiration to solve them!

As you work through these puzzles, you will feel your brain "waking up." I hope you will enjoy area mazes. There is a delightful sense of achievement that comes with saying "I got it!"

—RYOICHI MURAKAMI, director of El Camino

HOW TO SOLVE AREA MAZES

Using the given lengths and areas, find the value of ⑦. Remember, the formula for the area of a rectangle is height × width.

   If your calculation creates a fraction or decimal, STOP and look for another way. Area mazes can be solved using whole numbers only!*

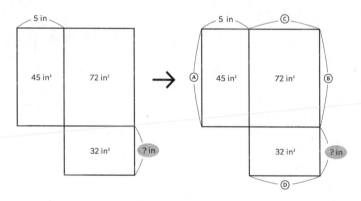

EXAMPLE ONE

Find length Ⓐ . . . 45 ÷ 5 = 9 in.

Find length Ⓑ . . . This is the same as Ⓐ, so 9 in.

Find length Ⓒ . . . 72 ÷ 9 = 8 in.

Find length Ⓓ . . . This is the same as Ⓒ, so 8 in.

Length ⑦ is 32 ÷ Ⓓ = 4 in.

*However, do not assume that *every* length or area in the puzzle must be a whole number.

EXAMPLE TWO

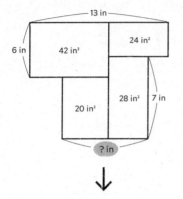

Note that the figures are not drawn to scale. You can't solve by "eyeballing"--you have to prove it with math!

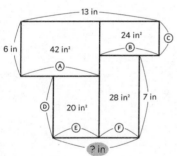

Even after you have solved a problem, you can revisit it to look for a more elegant solution.

Find length Ⓐ . . . $42 \div 6 = 7$ in.
Find length Ⓑ . . . $13 - 7 = 6$ in.
Find length Ⓒ . . . $24 \div 6 = 4$ in.
Find length Ⓓ . . . $(4 + 7) - 6 = 5$ in.
Find length Ⓔ . . . $20 \div 5 = 4$ in.
Find length Ⓕ . . . $28 \div 7 = 4$ in.
Length ⑦ is Ⓔ + Ⓕ = 8 in.

EXAMPLE THREE

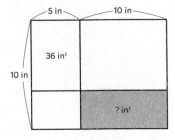

First, consider the two rectangles on the left together.

↓

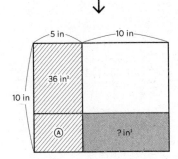

Find the area with stripes:

5 × 10 = 50 in.²

Find area Ⓐ . . . 50 − 36 = 14 in.²

Next, we will look at the two rectangles on the bottom.

↓

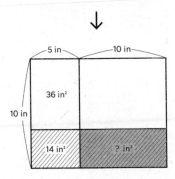

Area ⑦ is the same height as area Ⓐ, and exactly twice as wide.

So, area ⑦ must be exactly double area Ⓐ.

Area ⑦ is Ⓐ × 2 = 28 in.²

EXAMPLE FOUR

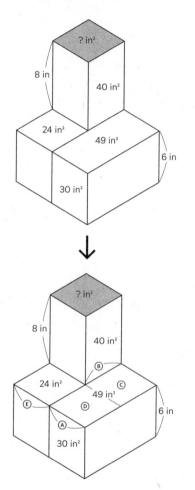

Some puzzles are 3D, but you can solve them using the same principles as the 2D puzzles.

Find length Ⓐ . . . 30 ÷ 6 = 5 in.

Find length Ⓑ . . . 40 ÷ 8 = 5 in.

Find area Ⓒ . . . 5 × 5 = 25 in.²

Find area Ⓓ . . . 49 − 25 = 24 in.²

Ⓓ equals the given area of 24, and they share a side.

Therefore length Ⓔ must equal length Ⓐ: Ⓔ = 5 in.

Area ⑦ is Ⓑ × Ⓔ = 25 in.²

PUZZLES

The puzzles get more challenging as you go.
This key will guide you!

LEVEL 1

LEVEL 2

LEVEL 3

LEVEL 4

LEVEL 5

PUZZLE 1

Find the solution on page 310.

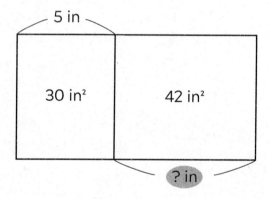

5 in

30 in²

42 in²

? in

SOLUTION

PUZZLE 2

Find the solution on page 310.

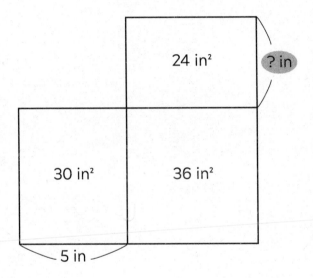

24 in²

? in

30 in²

36 in²

5 in

SOLUTION

PUZZLE 3

Find the solution on page 310.

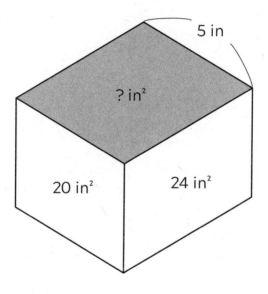

5 in

? in²

20 in²

24 in²

SOLUTION

PUZZLE 4

Find the solution on page 310.

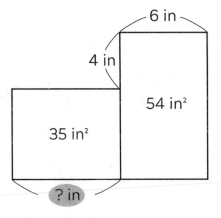

6 in

4 in

54 in²

35 in²

? in

SOLUTION

PUZZLE 5

Find the solution on page 310.

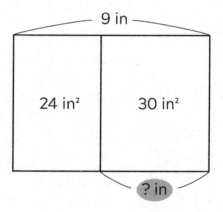

9 in

24 in²   30 in²

? in

SOLUTION

PUZZLE 6

Find the solution on page 311.

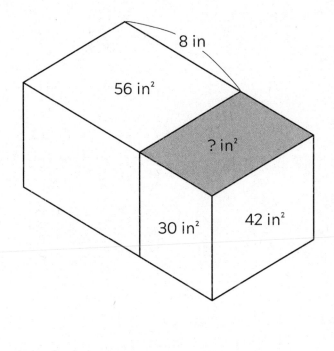

8 in

56 in²

? in²

30 in²

42 in²

SOLUTION

PUZZLE 7

Find the solution on page 311.

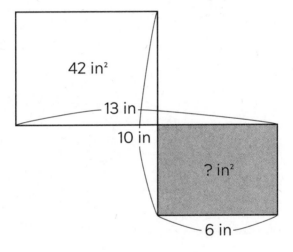

42 in²

13 in

10 in

? in²

6 in

SOLUTION

PUZZLE 8

Find the solution on page 311.

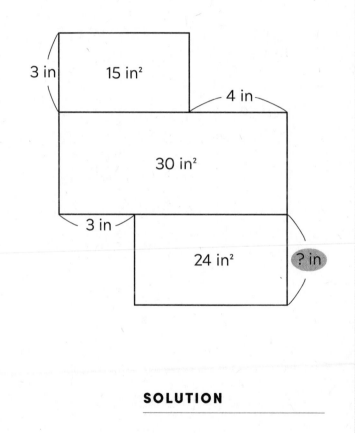

3 in   15 in²

4 in

30 in²

3 in

24 in²   ? in

SOLUTION

PUZZLE 9

Find the solution on page 311.

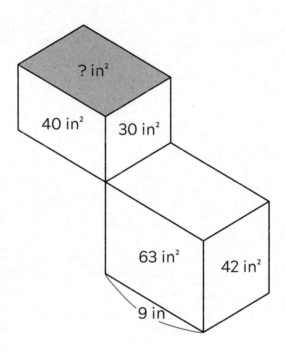

? in²

40 in²

30 in²

63 in²

42 in²

9 in

SOLUTION

PUZZLE 10

Find the solution on page 312.

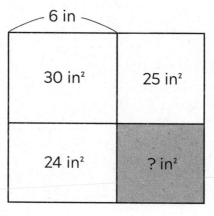

SOLUTION

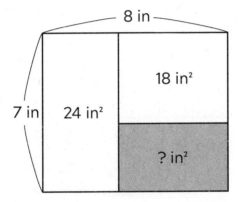

SOLUTION

PUZZLE 12

Find the solution on page 312.

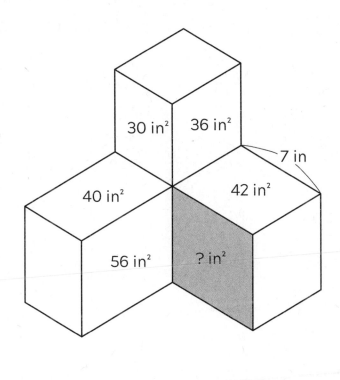

30 in² 36 in²

7 in

40 in² 42 in²

56 in² ? in²

SOLUTION

_____

PUZZLE 13

Find the solution on page 312.

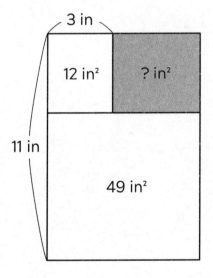

SOLUTION

PUZZLE 14

Find the solution on page 313.

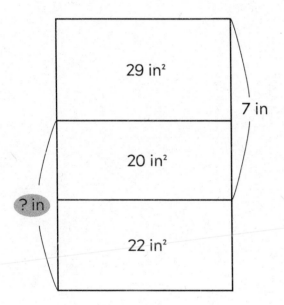

SOLUTION

_____

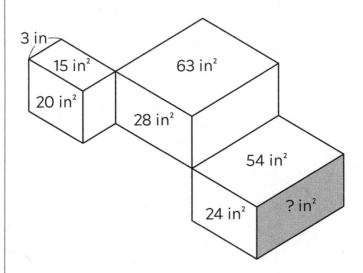

3 in

15 in²

20 in²

63 in²

28 in²

54 in²

24 in²

? in²

SOLUTION

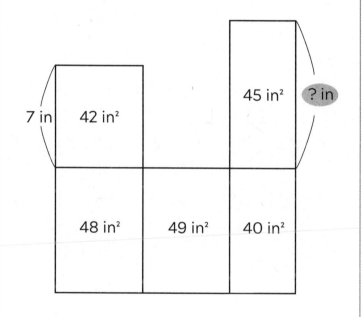

45 in² ? in

7 in 42 in²

48 in² 49 in² 40 in²

SOLUTION

PUZZLE 17

Find the solution on page 313.

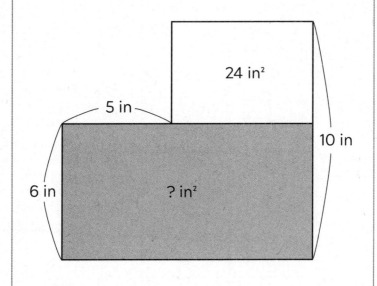

SOLUTION

PUZZLE 18

Find the solution on page 314.

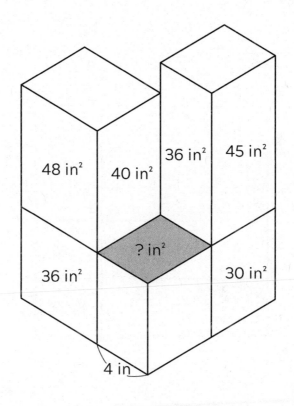

48 in²   40 in²   36 in²   45 in²

? in²

36 in²   30 in²

4 in

SOLUTION

PUZZLE 19

Find the solution on page 314.

5 in

35 in²		28 in²
	27 in²	
? in²		16 in²

SOLUTION

PUZZLE 20

Find the solution on page 314.

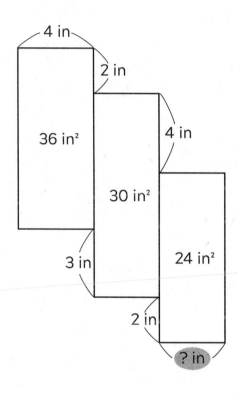

4 in

2 in

36 in²

4 in

30 in²

24 in²

3 in

2 in

? in

SOLUTION

PUZZLE 21

Find the solution on page 314.

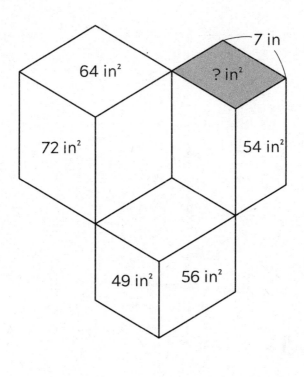

64 in²

7 in

? in²

72 in²

54 in²

49 in²   56 in²

SOLUTION

PUZZLE 22

Find the solution on page 315.

	20 in²	24 in²
3 in — 12 in²	15 in²	
32 in²		? in²

SOLUTION

PUZZLE 23

Find the solution on page 315.

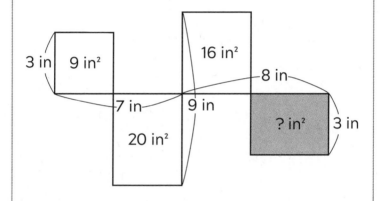

SOLUTION

PUZZLE 24

Find the solution on page 315.

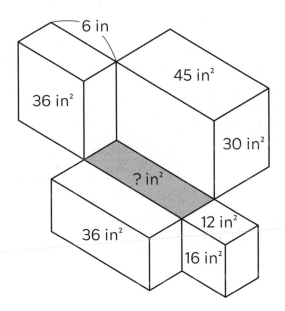

6 in

45 in²

36 in²

30 in²

? in²

36 in²

12 in²

16 in²

SOLUTION

PUZZLE 25

Find the solution on page 315.

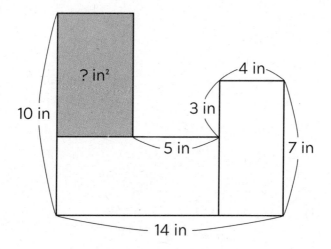

SOLUTION

PUZZLE 26

Find the solution on page 316.

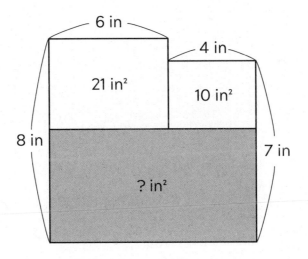

6 in

4 in

21 in²

10 in²

8 in

7 in

? in²

SOLUTION

PUZZLE 27

Find the solution on page 316.

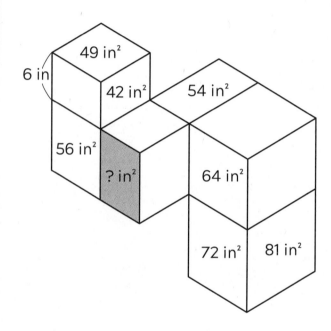

SOLUTION

PUZZLE 28

Find the solution on page 316.

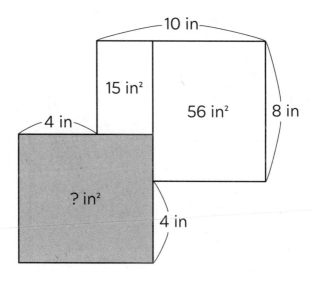

10 in

15 in²

56 in²

8 in

4 in

? in²

4 in

SOLUTION

PUZZLE 29

Find the solution on page 316.

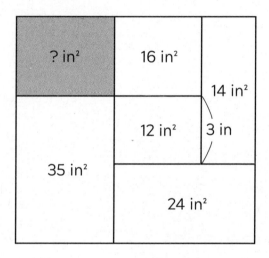

? in²   16 in²

14 in²

12 in²   3 in

35 in²

24 in²

SOLUTION

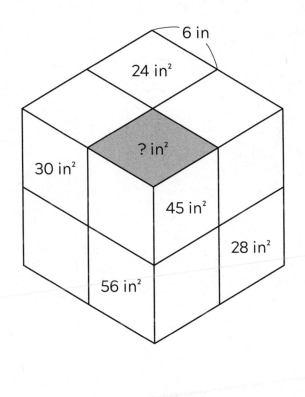

6 in

24 in²

? in²

30 in²

45 in²

28 in²

56 in²

SOLUTION

_____

PUZZLE 31

Find the solution on page 317.

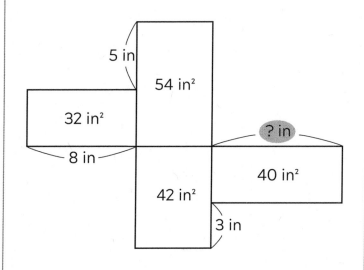

SOLUTION

PUZZLE 32

Find the solution on page 317.

6 in

| 20 in² | | 27 in² | | 24 in² |

| 25 in² | | ? in² | | 32 in² |

| 42 in² | | 36 in² |

SOLUTION

PUZZLE 33

Find the solution on page 317.

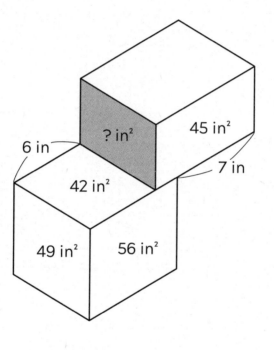

SOLUTION

PUZZLE 34

Find the solution on page 318.

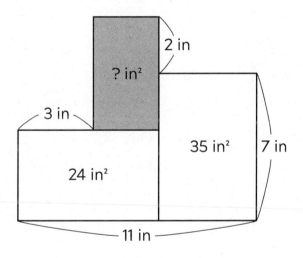

? in²

2 in

3 in

24 in²

35 in²

7 in

11 in

SOLUTION

PUZZLE 35

Find the solution on page 318.

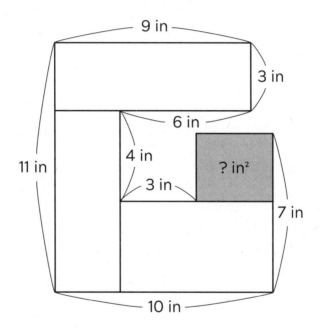

9 in

3 in

6 in

4 in

3 in

? in²

11 in

7 in

10 in

SOLUTION

PUZZLE 36

Find the solution on page 318.

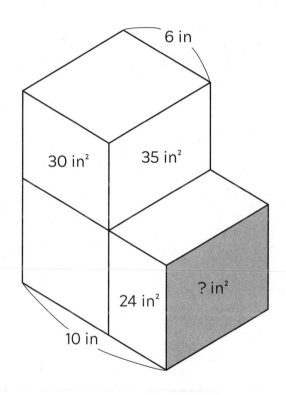

6 in

30 in²   35 in²

24 in²   ? in²

10 in

SOLUTION

PUZZLE 37

Find the solution on page 318.

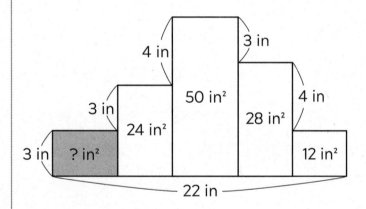

SOLUTION

PUZZLE 38

Find the solution on page 319.

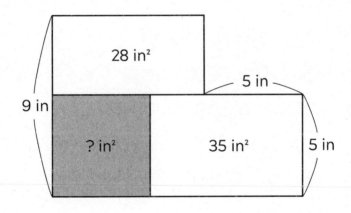

28 in²

5 in

9 in

? in²

35 in²

5 in

SOLUTION

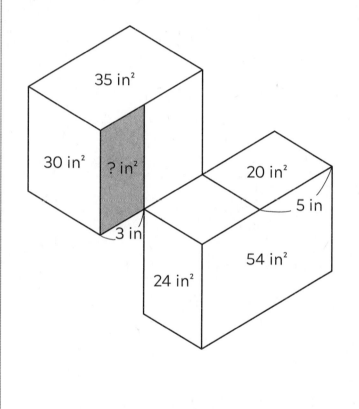

35 in²

30 in²

? in²

20 in²

5 in

3 in

54 in²

24 in²

SOLUTION

PUZZLE 40

Find the solution on page 319.

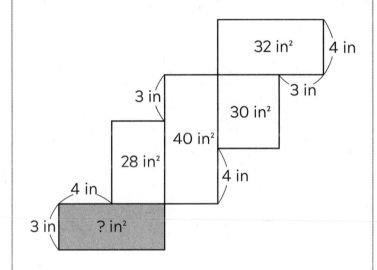

32 in² 4 in

3 in 3 in

30 in²

40 in²

28 in² 4 in

4 in

3 in ? in²

SOLUTION

PUZZLE 41

Find the solution on page 319.

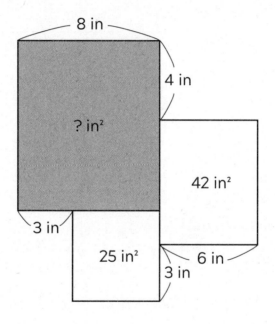

8 in

4 in

? in²

42 in²

3 in

25 in²

3 in

6 in

SOLUTION

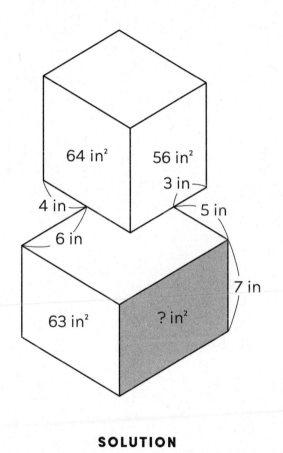

64 in²   56 in²

3 in

4 in

5 in

6 in

7 in

63 in²   ? in²

SOLUTION

_____

PUZZLE 43

Find the solution on page 320.

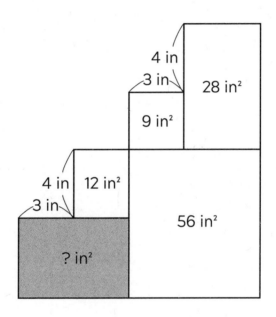

4 in

3 in

28 in²

9 in²

4 in

12 in²

3 in

56 in²

? in²

SOLUTION

PUZZLE 44

Find the solution on page 320.

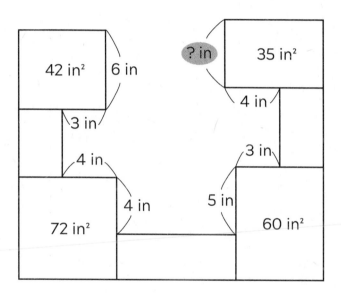

42 in² | 6 in

3 in

4 in

72 in²

4 in

? in | 35 in²

4 in

3 in

5 in

60 in²

SOLUTION

_____

PUZZLE 45

Find the solution on page 321.

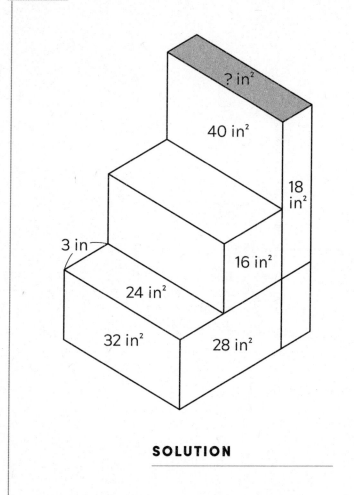

? in²

40 in²

18 in²

3 in

16 in²

24 in²

32 in²

28 in²

SOLUTION

PUZZLE 46

Find the solution on page 321.

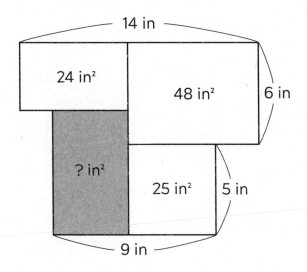

14 in

24 in²

48 in²

6 in

? in²

25 in²

5 in

9 in

SOLUTION

PUZZLE 47

Find the solution on page 321.

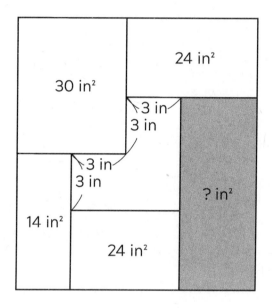

SOLUTION

PUZZLE 48

Find the solution on page 322.

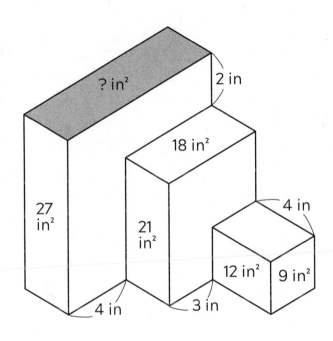

? in²

2 in

18 in²

27 in²

21 in²

4 in

4 in

12 in²   9 in²

3 in

SOLUTION

PUZZLE 49

Find the solution on page 322.

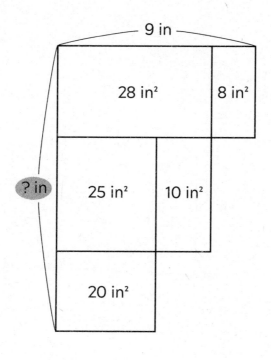

SOLUTION

_____

PUZZLE 50

Find the solution on page 322.

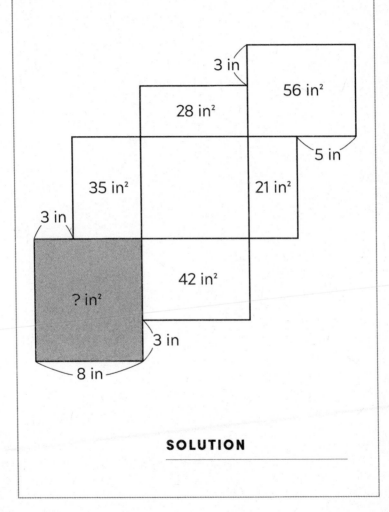

3 in

56 in²

28 in²

5 in

35 in²

21 in²

3 in

42 in²

? in²

3 in

8 in

SOLUTION

PUZZLE 51

Find the solution on page 323.

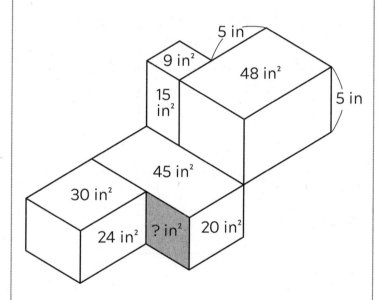

SOLUTION

PUZZLE 52

Find the solution on page 323.

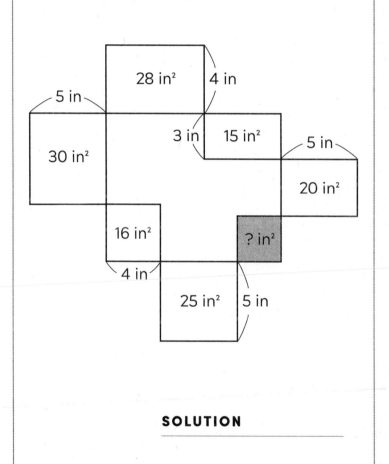

28 in² 4 in

5 in

30 in²

3 in 15 in² 5 in

20 in²

16 in²

? in²

4 in

25 in² 5 in

SOLUTION

PUZZLE 53

Find the solution on page 323.

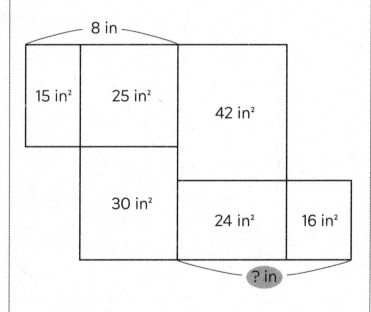

SOLUTION

___

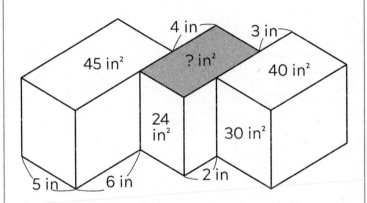

4 in   3 in

45 in²

? in²

40 in²

24 in²

30 in²

5 in   6 in   2 in

SOLUTION

PUZZLE 55

Find the solution on page 324.

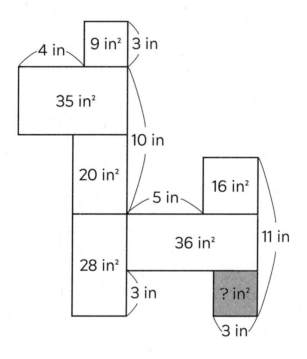

SOLUTION

PUZZLE 56

Find the solution on page 324.

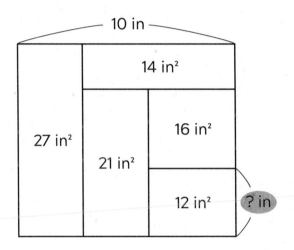

SOLUTION

PUZZLE 57

Find the solution on page 325.

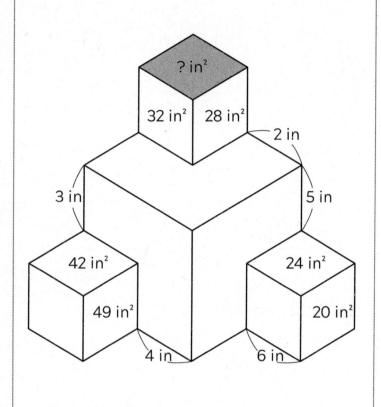

SOLUTION

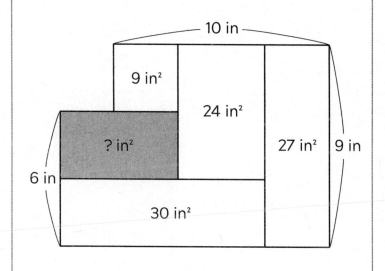

SOLUTION

Find the solution on page 325.

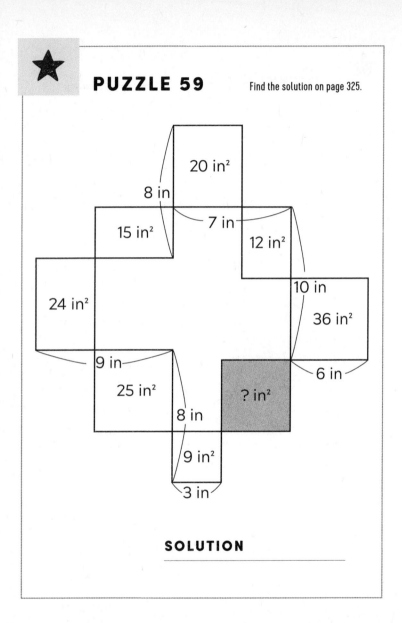

SOLUTION

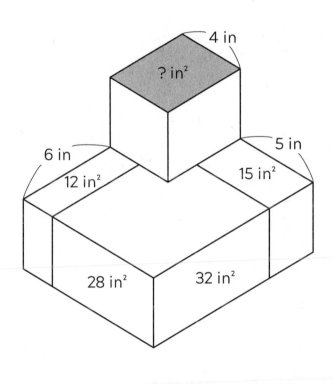

SOLUTION

PUZZLE 61

Find the solution on page 326.

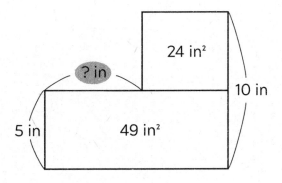

24 in²

? in

10 in

5 in

49 in²

SOLUTION

PUZZLE 62

Find the solution on page 326.

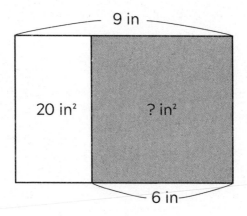

9 in

20 in² ? in²

6 in

SOLUTION

PUZZLE 63

Find the solution on page 326.

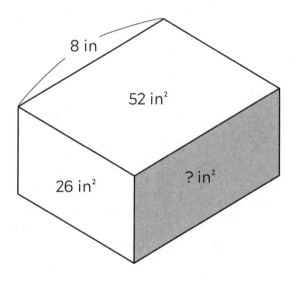

8 in

52 in²

26 in²

? in²

SOLUTION

Find the solution on page 327.

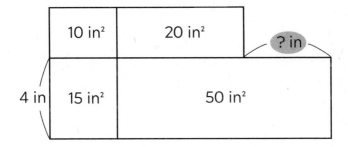

10 in²   20 in²

? in

4 in   15 in²   50 in²

SOLUTION

PUZZLE 65

Find the solution on page 327.

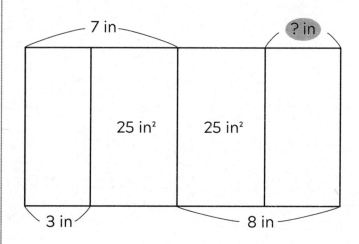

SOLUTION

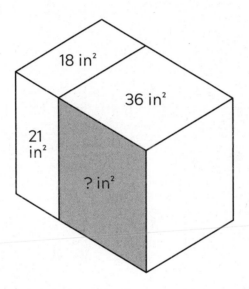

18 in²

36 in²

21 in²

? in²

SOLUTION

PUZZLE 67

Find the solution on page 327.

19 in²	38 in²	
21 in²		30 in²
	14 in²	? in²

SOLUTION

PUZZLE 68

Find the solution on page 328.

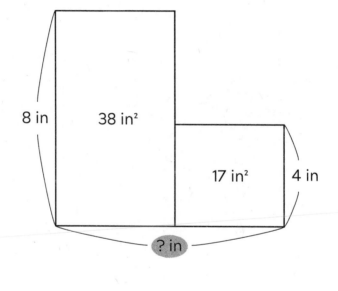

8 in   38 in²

17 in²   4 in

? in

SOLUTION

PUZZLE 69

Find the solution on page 328.

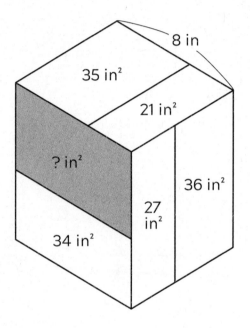

8 in

35 in²

21 in²

? in²

36 in²

27 in²

34 in²

SOLUTION

PUZZLE 70

Find the solution on page 328.

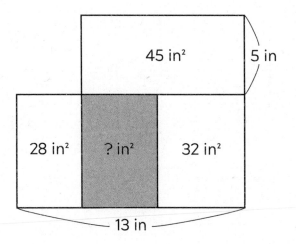

45 in²

5 in

28 in²   ? in²   32 in²

13 in

SOLUTION

PUZZLE 71

Find the solution on page 328.

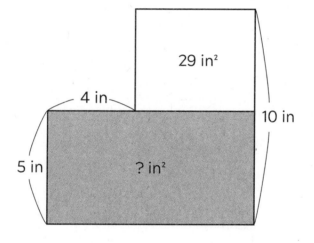

29 in²

4 in

10 in

5 in

? in²

SOLUTION

Find the solution on page 329.

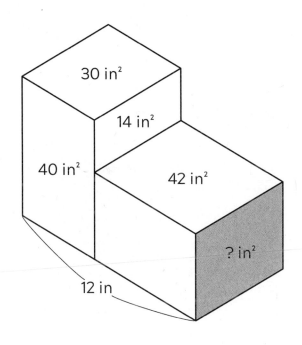

30 in²

14 in²

40 in²

42 in²

? in²

12 in

SOLUTION

PUZZLE 73

Find the solution on page 329.

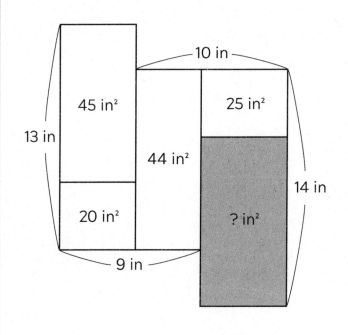

SOLUTION

PUZZLE 74

Find the solution on page 329.

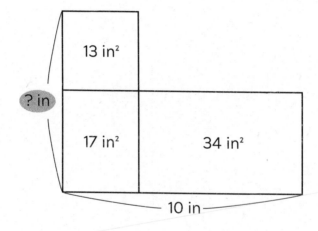

13 in²

? in

17 in²

34 in²

10 in

SOLUTION

PUZZLE 75

Find the solution on page 329.

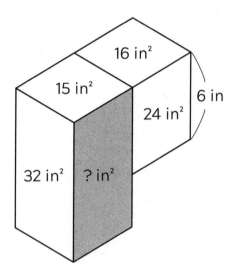

16 in²

15 in²

6 in

24 in²

32 in²

? in²

SOLUTION

PUZZLE 76

Find the solution on page 330.

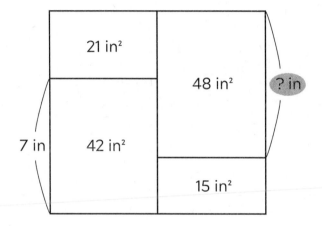

21 in²

48 in²

? in

7 in   42 in²

15 in²

SOLUTION

PUZZLE 77

Find the solution on page 330.

25 in²	20 in²	
30 in²		? in²
4 in 16 in²		24 in²
	45 in²	36 in²

SOLUTION

PUZZLE 78

Find the solution on page 330.

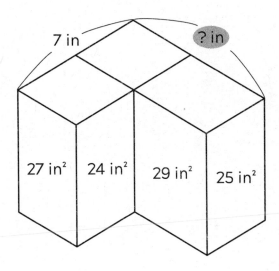

7 in

? in

27 in² 24 in² 29 in² 25 in²

SOLUTION

PUZZLE 79

Find the solution on page 330.

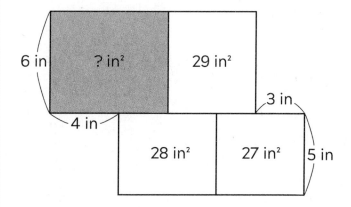

6 in

? in²

29 in²

3 in

4 in

28 in²

27 in²

5 in

SOLUTION

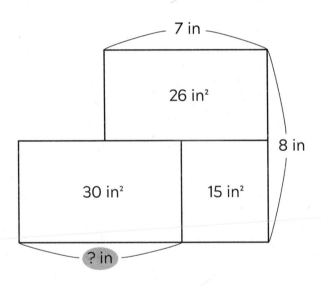

PUZZLE 81

Find the solution on page 331.

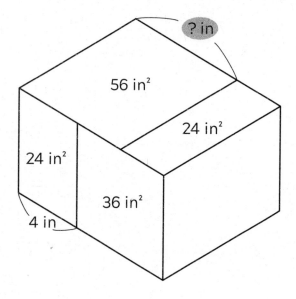

SOLUTION

PUZZLE 82

Find the solution on page 331.

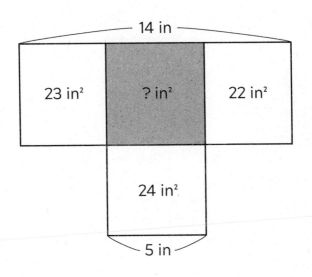

14 in

23 in²   ? in²   22 in²

24 in²

5 in

SOLUTION

? in²	56 in²	64 in²	72 in²

24 in²	9 in²	12 in²	13 in²	11 in²	14 in²	13 in²

SOLUTION

PUZZLE 84

Find the solution on page 332.

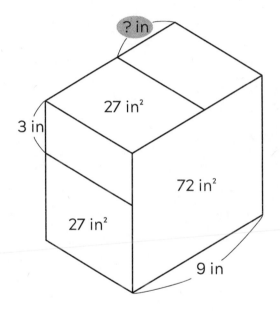

? in

27 in²

3 in

72 in²

27 in²

9 in

SOLUTION

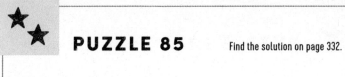
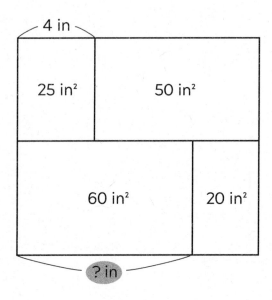

4 in

| 25 in² | 50 in² |
| 60 in² | 20 in² |

? in

SOLUTION

Find the solution on page 332.

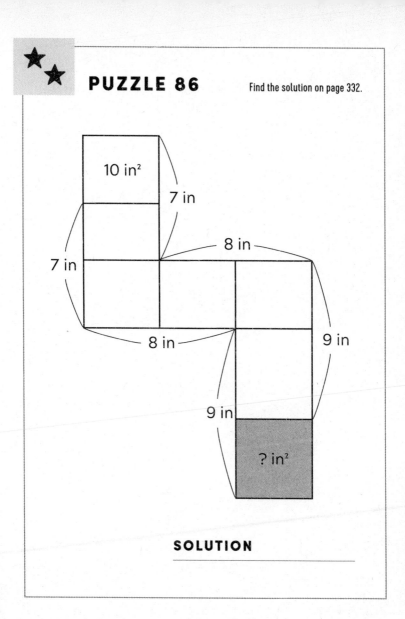

10 in²

7 in

8 in

7 in

8 in

9 in

9 in

? in²

SOLUTION

PUZZLE 87

Find the solution on page 332.

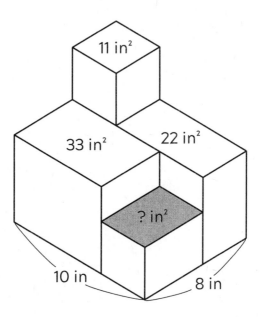

SOLUTION

PUZZLE 88

Find the solution on page 333.

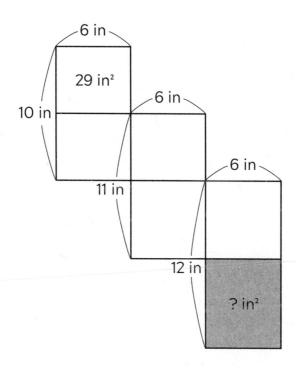

6 in

29 in²

10 in

6 in

11 in

6 in

12 in

? in²

SOLUTION

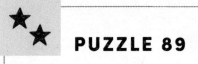

PUZZLE 89

Find the solution on page 333.

12 in²	24 in²	
15 in²		27 in²
	10 in²	
		14 in²

18 in²

? in²

SOLUTION

PUZZLE 90

Find the solution on page 333.

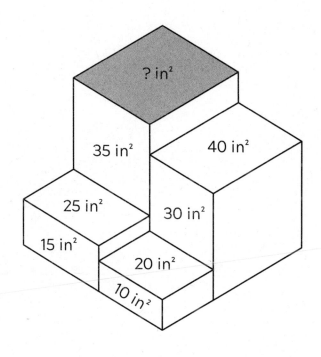

? in²

35 in²

40 in²

25 in²

30 in²

15 in²

20 in²

10 in²

SOLUTION

Find the solution on page 333.

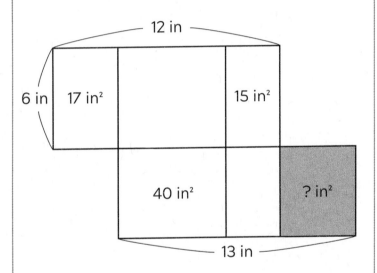

12 in

6 in

17 in²

15 in²

40 in²

? in²

13 in

SOLUTION

PUZZLE 92

Find the solution on page 334.

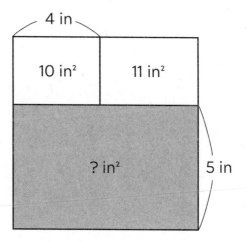

4 in

10 in²   11 in²

? in²   5 in

SOLUTION

PUZZLE 93

Find the solution on page 334.

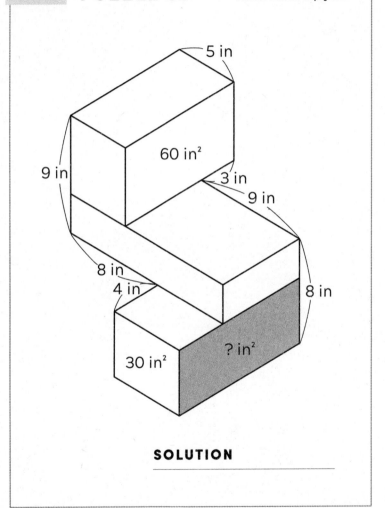

5 in

60 in²

9 in

3 in

9 in

8 in

4 in

8 in

30 in²

? in²

SOLUTION

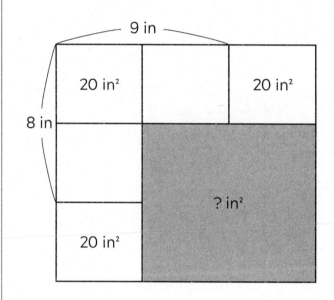

9 in

8 in

20 in²

20 in²

20 in²

? in²

SOLUTION

PUZZLE 95

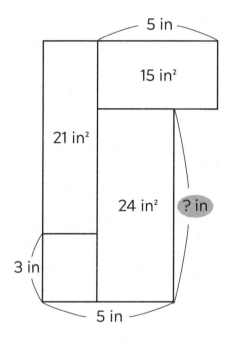

Find the solution on page 334.

5 in

15 in²

21 in²

24 in² ? in

3 in

5 in

SOLUTION

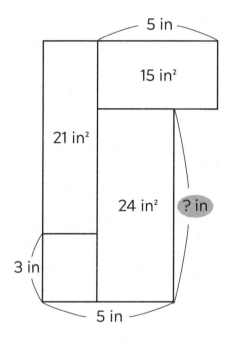

PUZZLE 96

Find the solution on page 335.

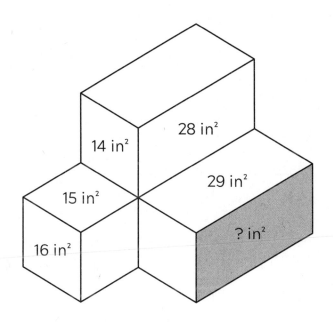

14 in²

28 in²

15 in²

29 in²

16 in²

? in²

SOLUTION

_____

PUZZLE 97

Find the solution on page 335.

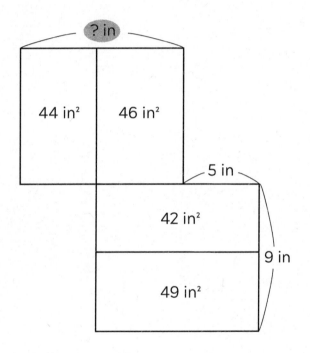

SOLUTION

104 ▸ THE BIG PUZZLE BOOK OF AREA MAZES

PUZZLE 98

Find the solution on page 335.

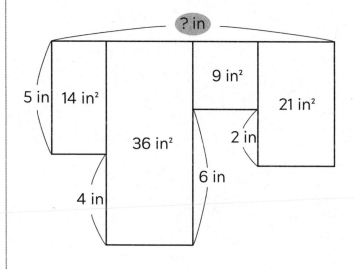

? in

9 in²

5 in 14 in²

21 in²

36 in²

2 in

6 in

4 in

SOLUTION

PUZZLE 99

Find the solution on page 336.

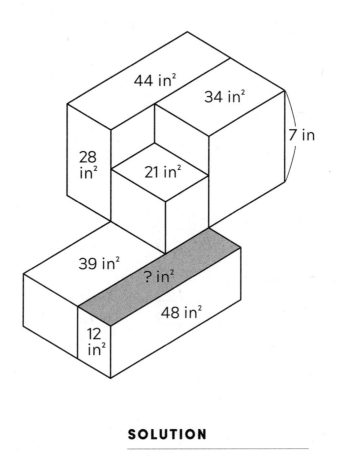

44 in²

34 in²

7 in

28 in²

21 in²

39 in²

? in²

48 in²

12 in²

SOLUTION

PUZZLE 100

Find the solution on page 336.

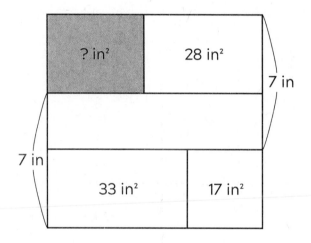

? in² 28 in² 7 in

7 in 33 in² 17 in²

SOLUTION

PUZZLE 101

Find the solution on page 336.

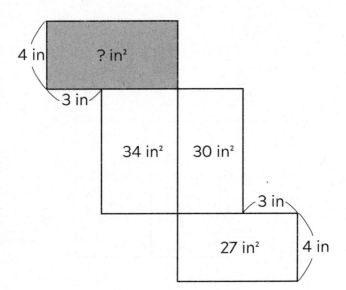

4 in

? in²

3 in

34 in²

30 in²

3 in

27 in²

4 in

SOLUTION

PUZZLE 102

Find the solution on page 336.

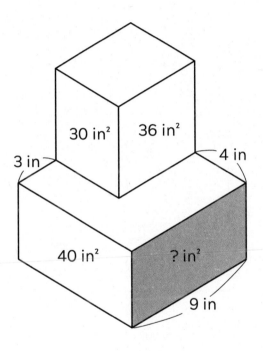

30 in²   36 in²

3 in   4 in

40 in²   ? in²

9 in

SOLUTION

PUZZLE 103

Find the solution on page 337.

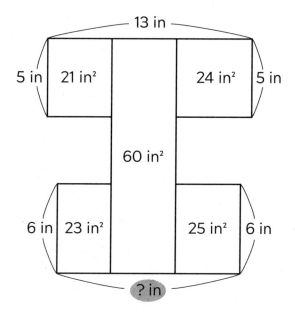

13 in

5 in | 21 in² | | 24 in² | 5 in

60 in²

6 in | 23 in² | | 25 in² | 6 in

? in

SOLUTION

PUZZLE 104

Find the solution on page 337.

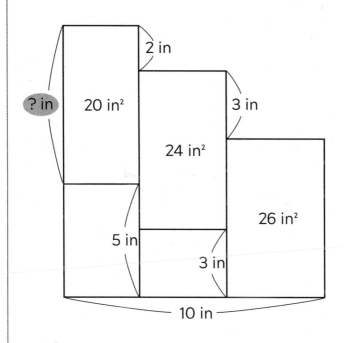

2 in

? in

20 in²

3 in

24 in²

5 in

26 in²

3 in

10 in

SOLUTION

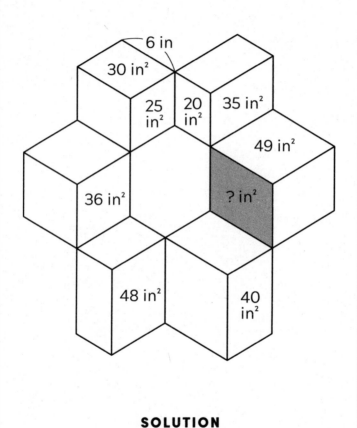

6 in

30 in²

25 in²

20 in²

35 in²

49 in²

36 in²

? in²

48 in²

40 in²

SOLUTION

PUZZLE 106

Find the solution on page 338.

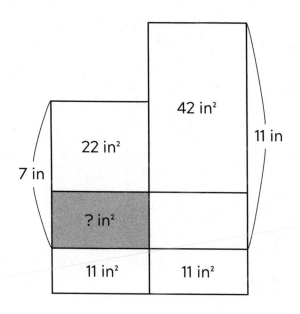

42 in²

22 in²

11 in

7 in

? in²

11 in²   11 in²

SOLUTION

PUZZLE 107

Find the solution on page 338.

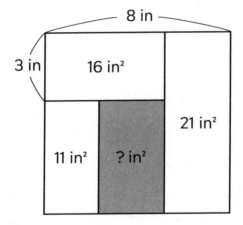

SOLUTION

PUZZLE 108

Find the solution on page 338.

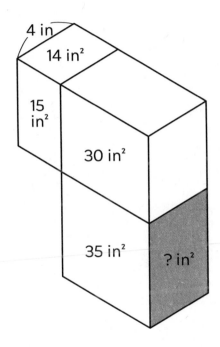

4 in

14 in²

15 in²

30 in²

35 in²

? in²

SOLUTION

PUZZLE 109

Find the solution on page 338.

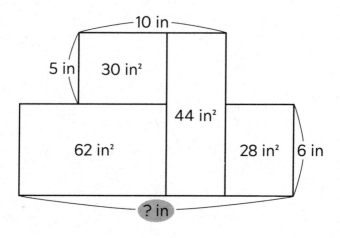

SOLUTION

PUZZLE 110

Find the solution on page 339.

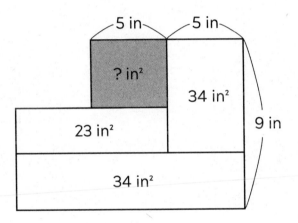

5 in   5 in

? in²

34 in²

23 in²

9 in

34 in²

SOLUTION

PUZZLE 111

Find the solution on page 339.

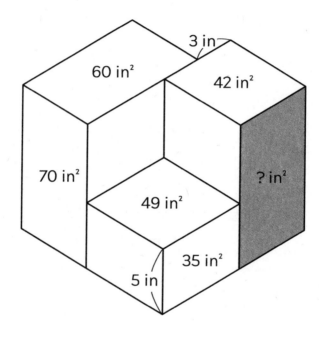

3 in

60 in²

42 in²

70 in²

49 in²

? in²

35 in²

5 in

SOLUTION

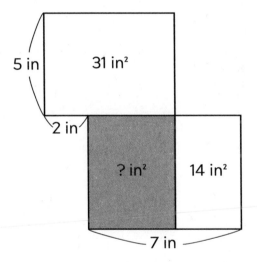

5 in

31 in²

2 in

? in²

14 in²

7 in

SOLUTION

PUZZLE 113

Find the solution on page 339.

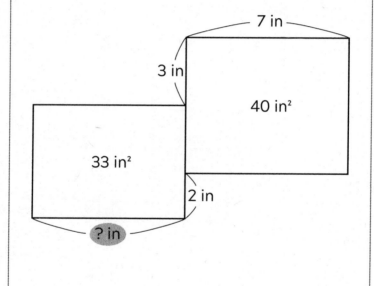

7 in

3 in

40 in²

33 in²

2 in

? in

SOLUTION

120 ▸ THE BIG PUZZLE BOOK OF AREA MAZES

Find the solution on page 340.

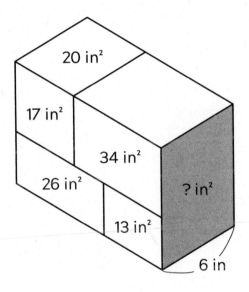

20 in²

17 in²

34 in²

26 in²

? in²

13 in²

6 in

SOLUTION

PUZZLE 115

Find the solution on page 340.

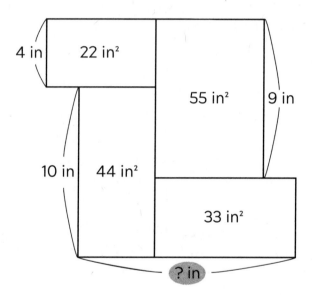

SOLUTION

PUZZLE 116

Find the solution on page 340.

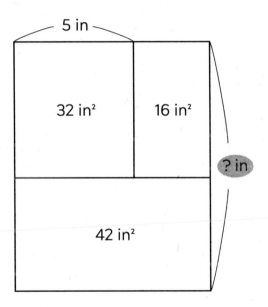

SOLUTION

PUZZLE 117

Find the solution on page 340.

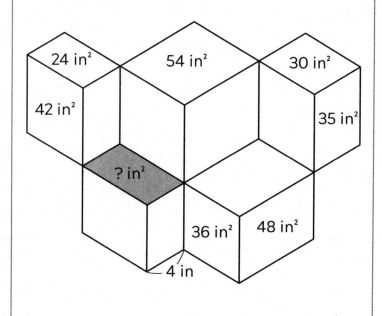

24 in²	54 in²	30 in²
42 in²		35 in²
? in²		
	36 in²	48 in²

4 in

SOLUTION

PUZZLE 118

Find the solution on page 341.

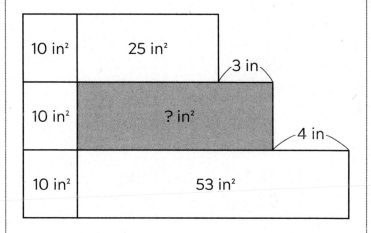

10 in²	25 in²	
10 in²	? in²	3 in
10 in²	53 in²	4 in

SOLUTION

_____

PUZZLE 119

Find the solution on page 341.

	35 in²		80 in²	
42 in²		25 in²		39 in²
	63 in²	? in²		
56 in²			64 in²	
				13 in²

SOLUTION

PUZZLE 120

Find the solution on page 341.

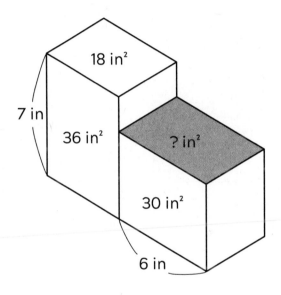

18 in²

7 in

36 in²

? in²

30 in²

6 in

SOLUTION

PUZZLE 121

Find the solution on page 341.

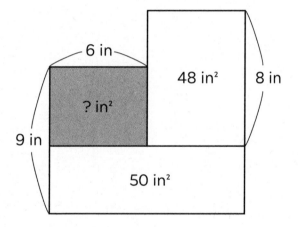

6 in

48 in²

8 in

? in²

9 in

50 in²

SOLUTION

PUZZLE 122

Find the solution on page 342.

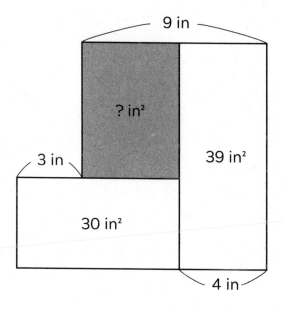

9 in

? in²

3 in

39 in²

30 in²

4 in

SOLUTION

PUZZLE 123

Find the solution on page 342.

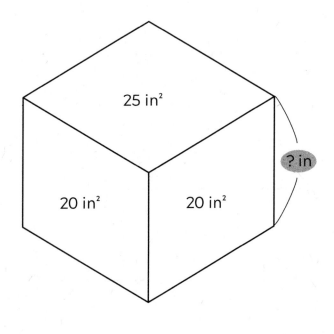

SOLUTION

PUZZLE 124

Find the solution on page 342.

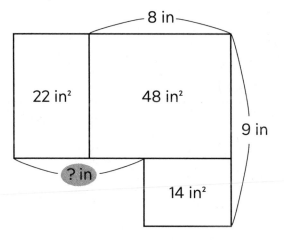

SOLUTION

PUZZLE 125

Find the solution on page 342.

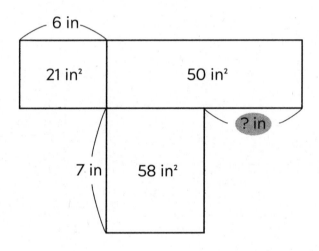

SOLUTION

PUZZLE 126

Find the solution on page 343.

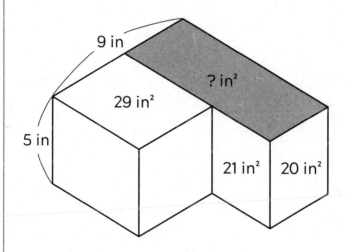

9 in

? in²

29 in²

5 in

21 in²   20 in²

SOLUTION

PUZZLE 127

Find the solution on page 343.

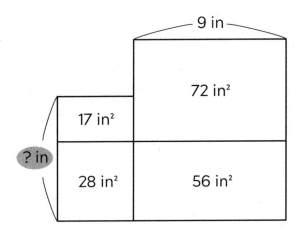

9 in

72 in²

17 in²

? in

28 in²

56 in²

SOLUTION

26 in²	? in²	52 in²
27 in²		80 in²
	20 in²	

SOLUTION

PUZZLE 129

Find the solution on page 343.

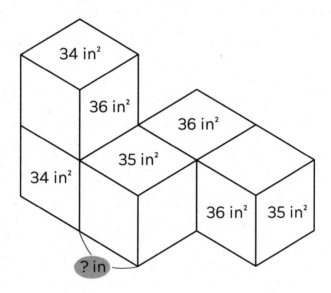

34 in²

36 in²

36 in²

35 in²

34 in²

36 in²   35 in²

? in

SOLUTION

PUZZLE 130

Find the solution on page 344.

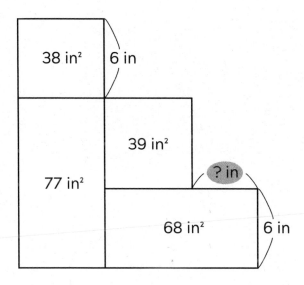

38 in²   6 in

39 in²

77 in²   ? in

68 in²   6 in

SOLUTION

PUZZLE 131

Find the solution on page 344.

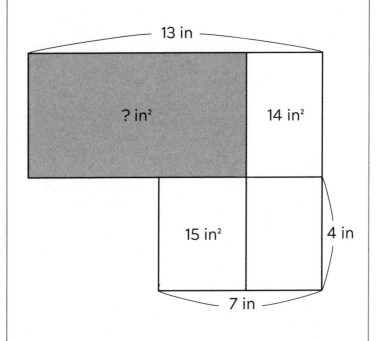

SOLUTION

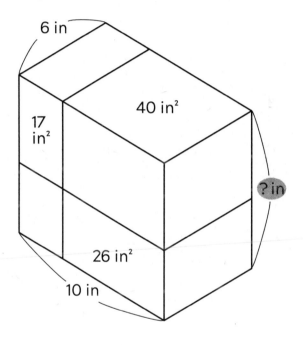

6 in

40 in²

17 in²

? in

26 in²

10 in

SOLUTION

PUZZLE 133

Find the solution on page 345.

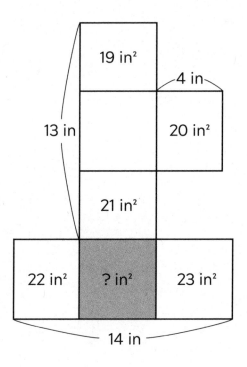

19 in²

4 in

13 in

20 in²

21 in²

22 in²

? in²

23 in²

14 in

SOLUTION

PUZZLE 134

Find the solution on page 345.

22 in²	11 in²
21 in²	12 in²

? in²	24 in²

SOLUTION

PUZZLE 135

Find the solution on page 345.

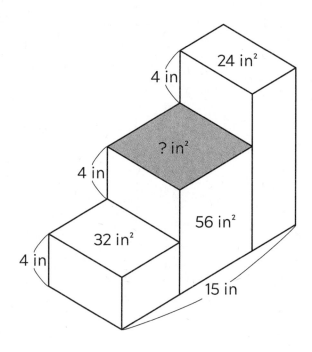

SOLUTION

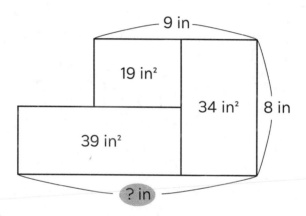

9 in

19 in²

34 in²   8 in

39 in²

? in

SOLUTION

PUZZLE 137

Find the solution on page 346.

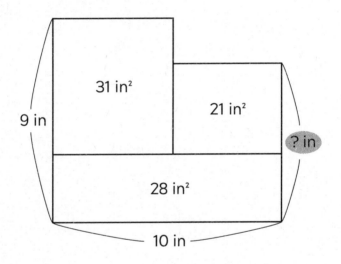

31 in²

21 in²

9 in

? in

28 in²

10 in

SOLUTION

PUZZLE 138

Find the solution on page 346.

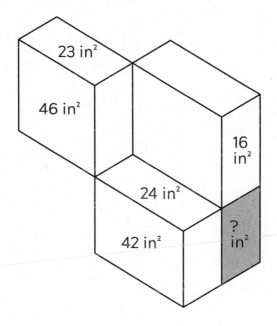

23 in²

46 in²

16 in²

24 in²

42 in²

? in²

SOLUTION

PUZZLE 139

Find the solution on page 346.

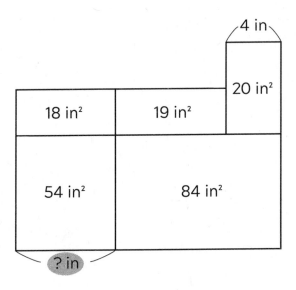

SOLUTION

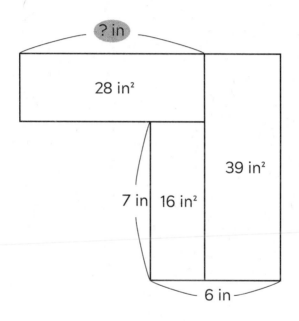

SOLUTION

PUZZLE 141

Find the solution on page 347.

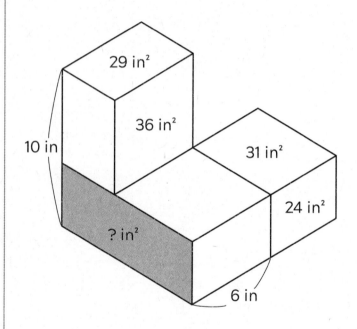

29 in²

36 in²

31 in²

10 in

24 in²

? in²

6 in

SOLUTION

PUZZLE 142

Find the solution on page 347.

	15 in²	
43 in²	? in²	35 in²
24 in²		30 in²

6 in

SOLUTION

PUZZLE 143

Find the solution on page 347.

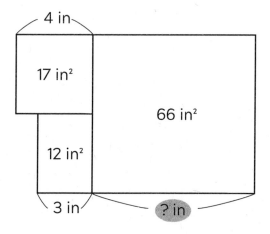

4 in

17 in²

66 in²

12 in²

3 in

? in

SOLUTION

PUZZLE 144

Find the solution on page 347.

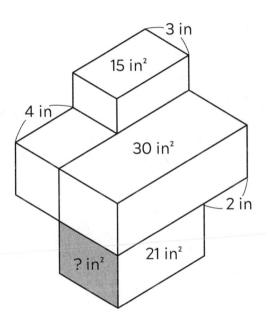

3 in

15 in²

4 in

30 in²

2 in

? in²

21 in²

SOLUTION

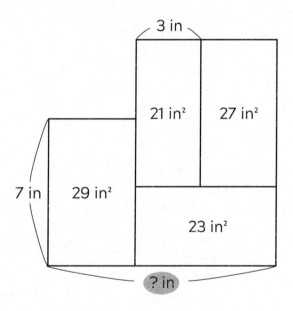

SOLUTION

PUZZLE 146

Find the solution on page 348.

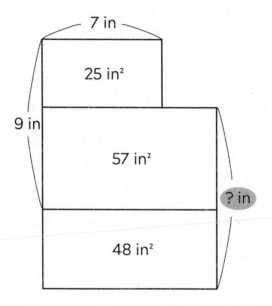

7 in

25 in²

9 in

57 in²

? in

48 in²

SOLUTION

PUZZLE 147

Find the solution on page 348.

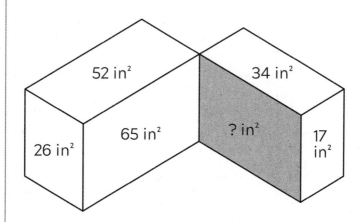

52 in²

34 in²

26 in²

65 in²

? in²

17 in²

SOLUTION

PUZZLE 148

Find the solution on page 348.

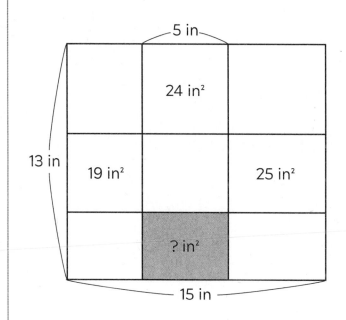

PUZZLE 149

Find the solution on page 349.

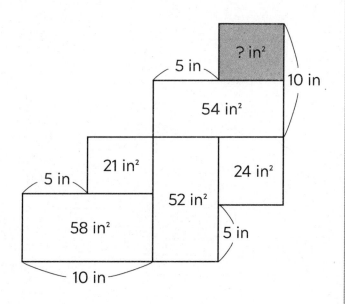

SOLUTION

PUZZLE 150

Find the solution on page 349.

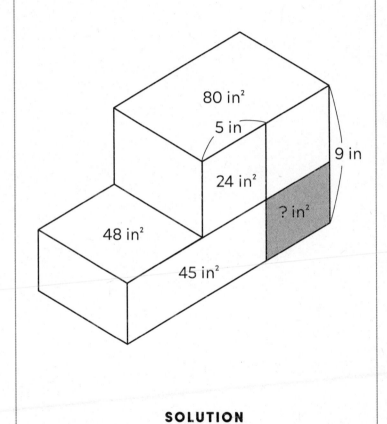

80 in²

5 in

9 in

24 in²

48 in²

? in²

45 in²

SOLUTION

PUZZLE 151

Find the solution on page 349.

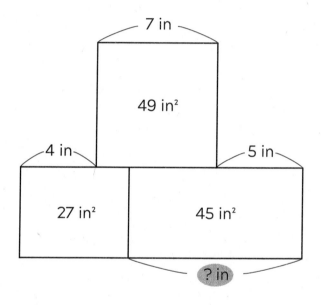

SOLUTION

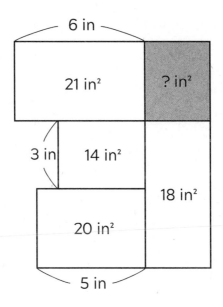

6 in

21 in²

? in²

3 in   14 in²

18 in²

20 in²

5 in

SOLUTION

PUZZLE 153

Find the solution on page 350.

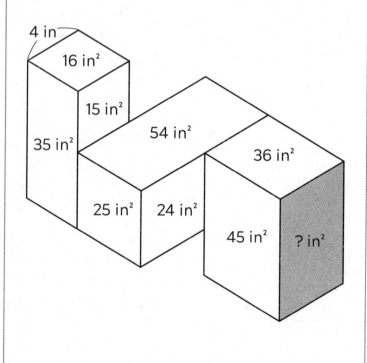

4 in

16 in²

15 in²

35 in²

54 in²

36 in²

25 in²

24 in²

45 in²

? in²

SOLUTION

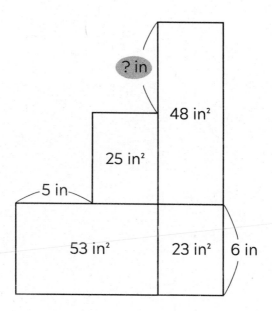

? in

48 in²

25 in²

5 in

53 in²   23 in²   6 in

SOLUTION

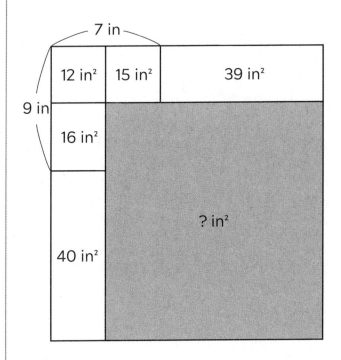

SOLUTION

PUZZLE 156

Find the solution on page 351.

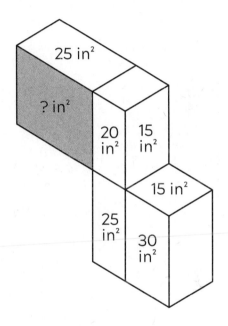

25 in²

? in²

20 in²

15 in²

15 in²

25 in²

30 in²

SOLUTION

PUZZLE 157

Find the solution on page 351.

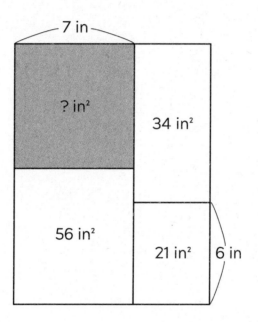

SOLUTION

PUZZLE 158

Find the solution on page 352.

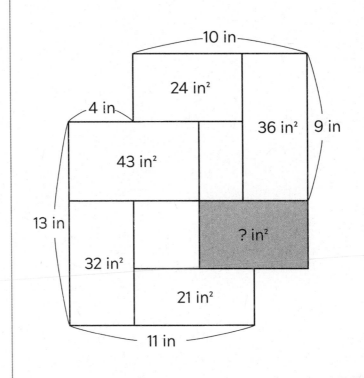

SOLUTION

PUZZLE 159

Find the solution on page 352.

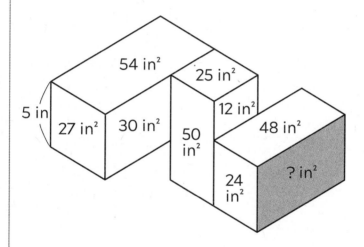

SOLUTION

PUZZLE 160

Find the solution on page 352.

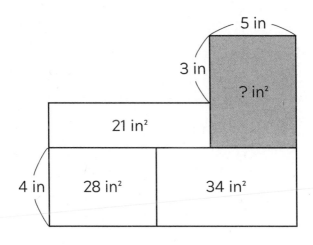

5 in

3 in

? in²

21 in²

4 in   28 in²   34 in²

SOLUTION

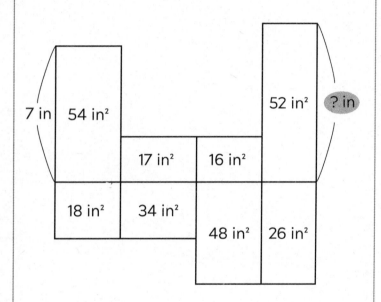

SOLUTION

PUZZLE 162

Find the solution on page 353.

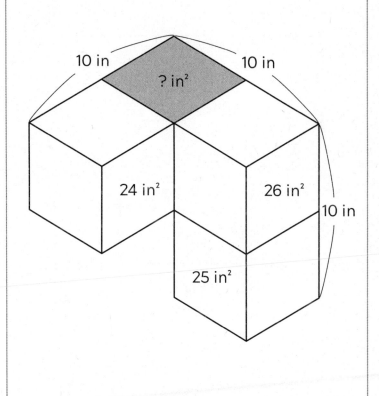

10 in

? in²

10 in

24 in²

26 in²

10 in

25 in²

SOLUTION

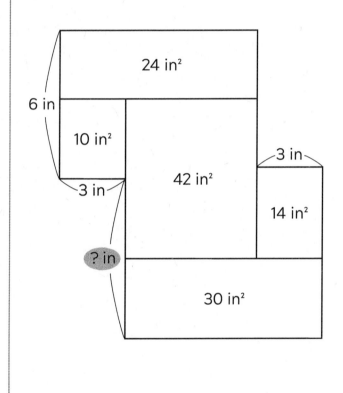

PUZZLE 164

Find the solution on page 354.

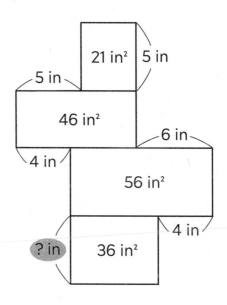

21 in² · 5 in

5 in

46 in²

6 in

4 in

56 in²

4 in

? in · 36 in²

SOLUTION

PUZZLE 165

Find the solution on page 354.

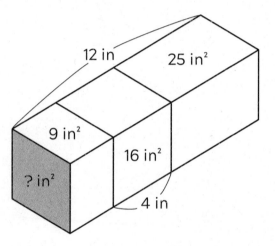

12 in

25 in²

9 in²

16 in²

4 in

? in²

SOLUTION

PUZZLE 166

Find the solution on page 354.

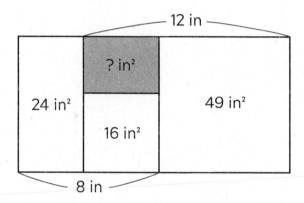

SOLUTION

Find the solution on page 354.

	16 in²	10 in²
30 in²		
		? in²
10 in²	12 in²	

SOLUTION

PUZZLE 168

Find the solution on page 355.

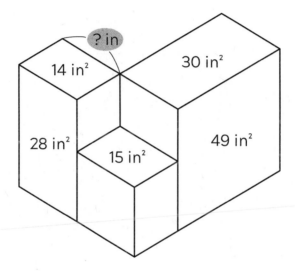

? in

14 in²

30 in²

28 in²

15 in²

49 in²

SOLUTION

PUZZLE 169

Find the solution on page 355.

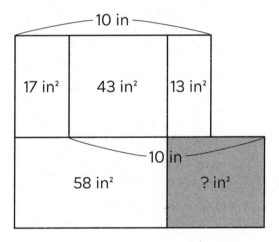

SOLUTION

PUZZLE 170

Find the solution on page 355.

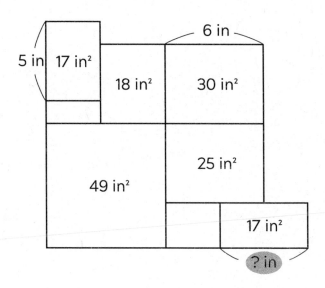

SOLUTION

_____

PUZZLE 171

Find the solution on page 355.

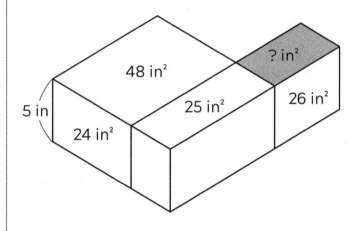

48 in²

? in²

5 in

24 in²

25 in²

26 in²

SOLUTION

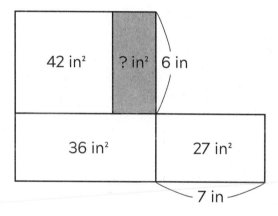

42 in² ? in² 6 in

36 in² 27 in²

7 in

SOLUTION

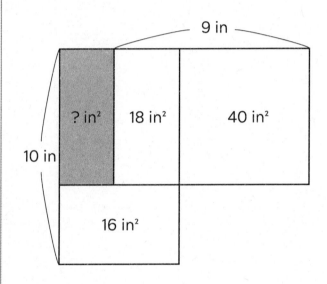

PUZZLE 174

Find the solution on page 356.

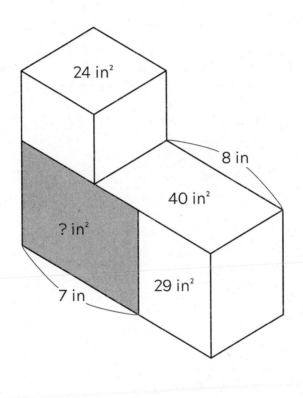

24 in²

8 in

40 in²

? in²

7 in

29 in²

SOLUTION

PUZZLE 175

Find the solution on page 356.

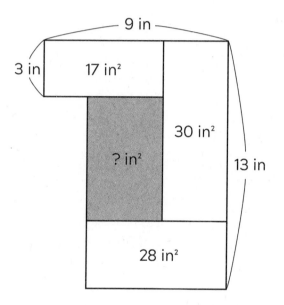

9 in

3 in

17 in²

30 in²

? in²

13 in

28 in²

SOLUTION

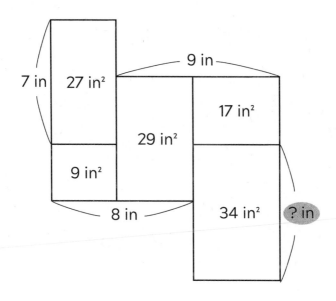

7 in | 27 in²

9 in

17 in²

29 in²

9 in²

8 in

34 in²

? in

SOLUTION

PUZZLE 177

Find the solution on page 357.

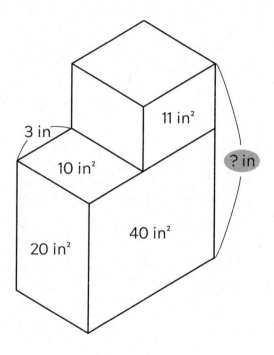

3 in

11 in²

10 in²

? in

20 in²

40 in²

SOLUTION

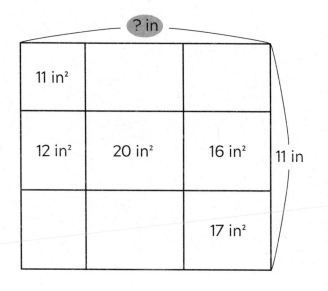

SOLUTION

PUZZLE 179

Find the solution on page 357.

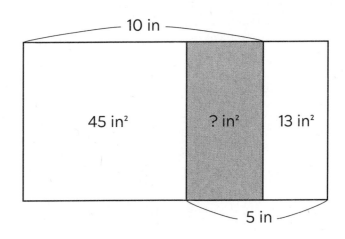

10 in

45 in²　　? in²　　13 in²

5 in

SOLUTION

PUZZLE 180

Find the solution on page 358.

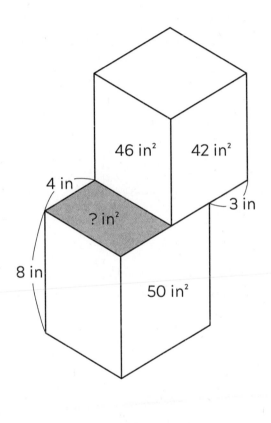

46 in² 42 in²

4 in

3 in

? in²

8 in

50 in²

SOLUTION

_____

PUZZLE 181

Find the solution on page 358.

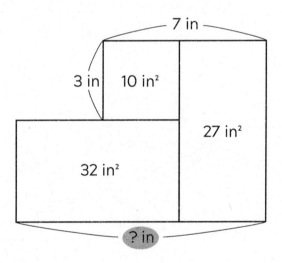

SOLUTION

PUZZLE 182

Find the solution on page 358.

38 in²	? in²
19 in²	12 in²
49 in²	24 in²

SOLUTION

PUZZLE 183

Find the solution on page 358.

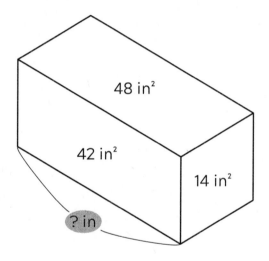

48 in²

42 in²

14 in²

? in

SOLUTION

PUZZLE 184

Find the solution on page 359.

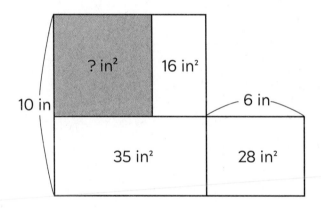

SOLUTION

PUZZLE 185

Find the solution on page 359.

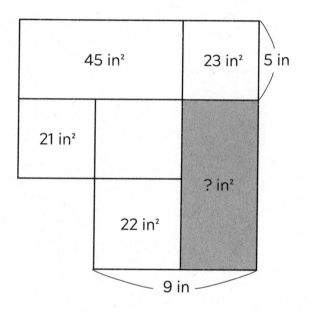

| 45 in² | 23 in² | 5 in |

| 21 in² |

| 22 in² | ? in² |

9 in

SOLUTION

PUZZLE 186

Find the solution on page 359.

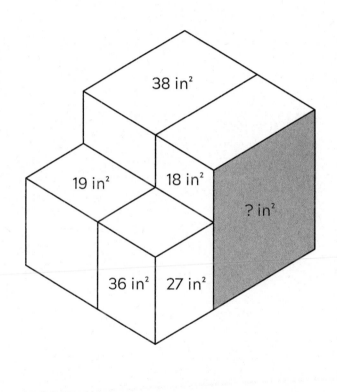

38 in²

19 in²

18 in²

? in²

36 in²   27 in²

SOLUTION

PUZZLE 187

Find the solution on page 359.

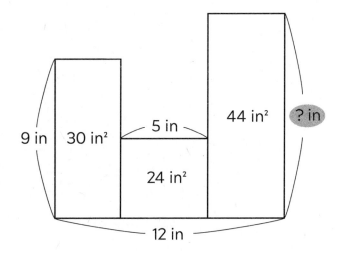

9 in | 30 in² | 5 in | 44 in² | ? in

24 in²

12 in

SOLUTION

PUZZLE 188

Find the solution on page 360.

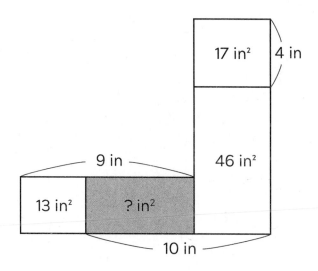

17 in² 4 in

9 in

46 in²

13 in² ? in²

10 in

SOLUTION

PUZZLE 189

Find the solution on page 360.

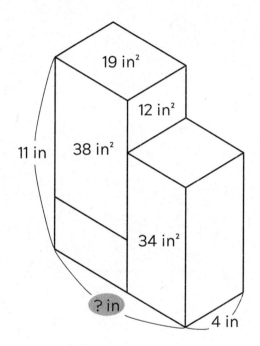

19 in²

12 in²

11 in

38 in²

34 in²

? in

4 in

SOLUTION

PUZZLE 190

Find the solution on page 360.

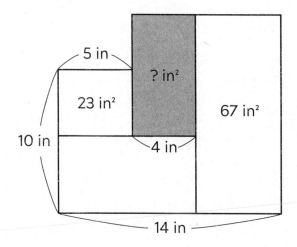

5 in

? in²

23 in²

67 in²

10 in

4 in

14 in

SOLUTION

PUZZLE 191

Find the solution on page 360.

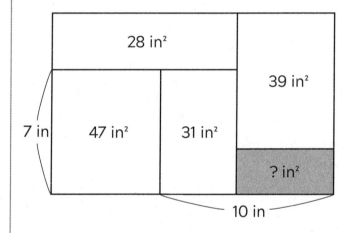

SOLUTION

PUZZLE 192

Find the solution on page 361.

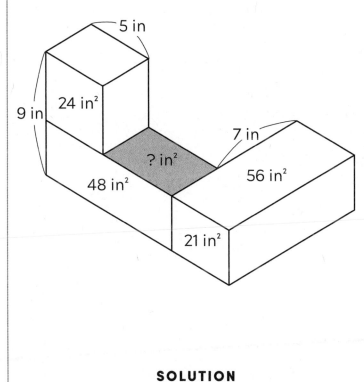

5 in

24 in²

9 in

7 in

? in²

48 in²

56 in²

21 in²

SOLUTION

PUZZLE 193

Find the solution on page 361.

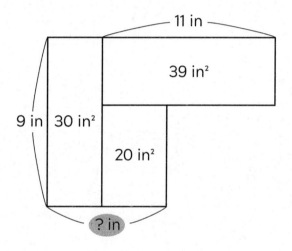

11 in

39 in²

9 in | 30 in²

20 in²

? in

SOLUTION

PUZZLE 194

Find the solution on page 361.

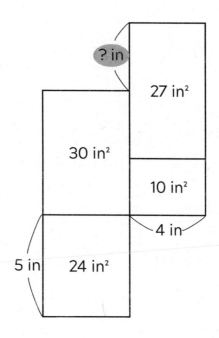

? in

27 in²

30 in²

10 in²

4 in

5 in   24 in²

SOLUTION

PUZZLE 195

Find the solution on page 361.

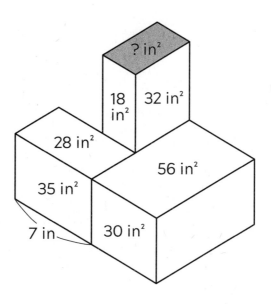

SOLUTION

PUZZLE 196

Find the solution on page 362.

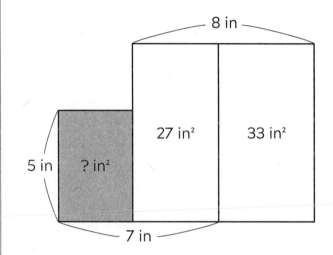

8 in

27 in²   33 in²

5 in   ? in²

7 in

SOLUTION

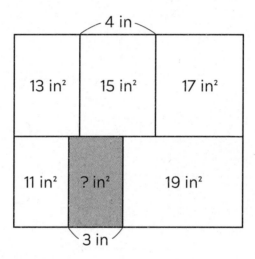

← 4 in →

13 in²	15 in²	17 in²

11 in²	? in²	19 in²

← 3 in →

SOLUTION

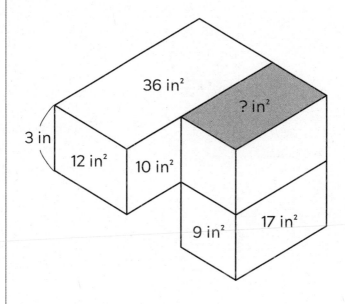

36 in²

? in²

3 in

12 in²

10 in²

9 in²

17 in²

SOLUTION

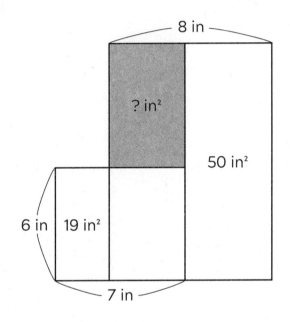

8 in

? in²

50 in²

6 in

19 in²

7 in

SOLUTION

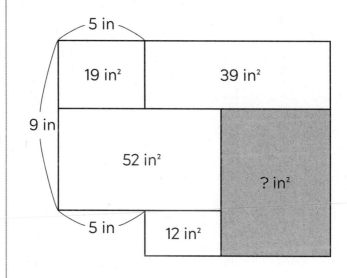

5 in

19 in²

39 in²

9 in

52 in²

? in²

5 in

12 in²

SOLUTION

PUZZLE 201

Find the solution on page 363.

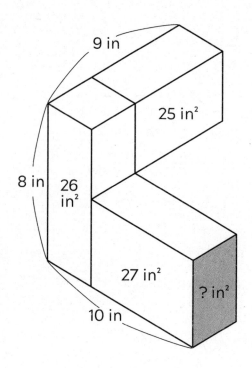

9 in

25 in²

8 in

26 in²

27 in²

? in²

10 in

SOLUTION

PUZZLE 202

Find the solution on page 363.

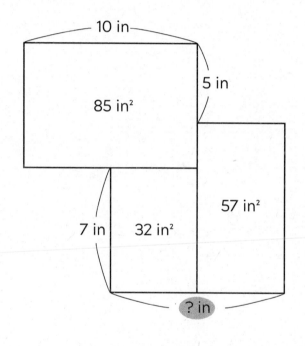

SOLUTION

PUZZLE 203

Find the solution on page 363.

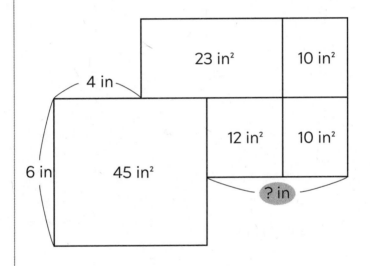

SOLUTION

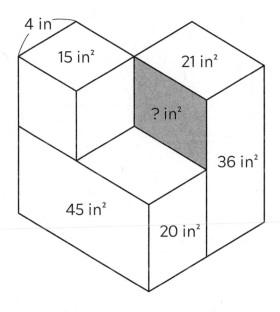

SOLUTION

PUZZLE 205

Find the solution on page 364.

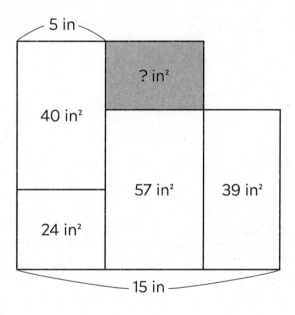

5 in

? in²

40 in²

57 in²

39 in²

24 in²

15 in

SOLUTION

PUZZLE 206

Find the solution on page 364.

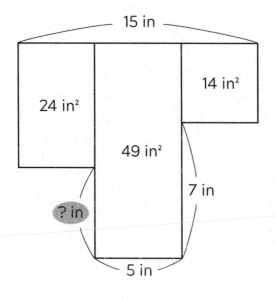

15 in

24 in²

14 in²

49 in²

7 in

? in

5 in

SOLUTION

PUZZLE 207

Find the solution on page 365.

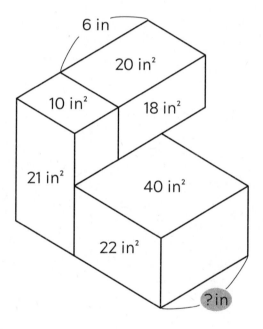

6 in

20 in²

10 in²

18 in²

21 in²

40 in²

22 in²

? in

SOLUTION

PUZZLE 208

Find the solution on page 365.

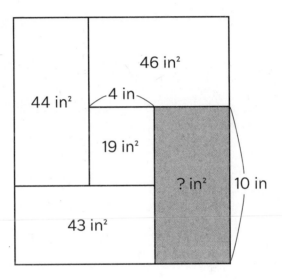

SOLUTION

PUZZLE 209

Find the solution on page 365.

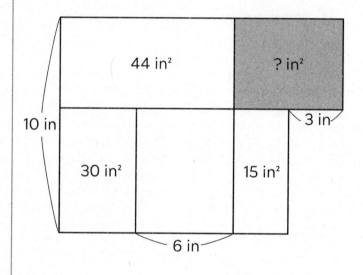

SOLUTION

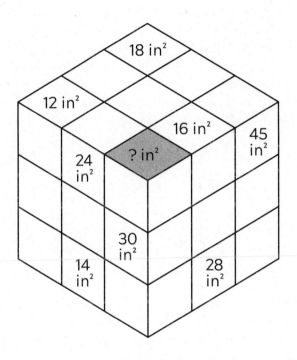

18 in²

12 in²

16 in²

45 in²

? in²

24 in²

30 in²

14 in²

28 in²

SOLUTION

PUZZLE 211

Find the solution on page 366.

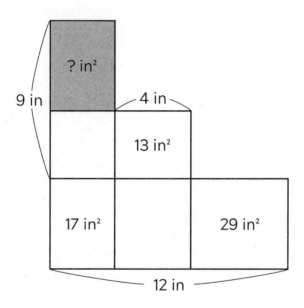

9 in

? in²

4 in

13 in²

17 in²

29 in²

12 in

SOLUTION

PUZZLE 212

Find the solution on page 366.

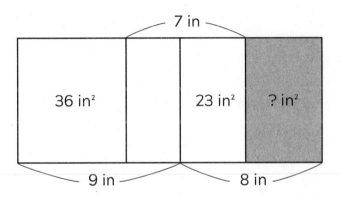

36 in²   23 in²   ? in²

7 in

9 in   8 in

SOLUTION

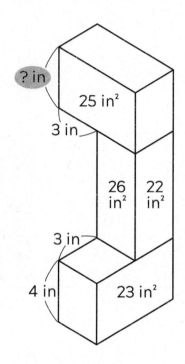

? in

25 in²

3 in

26 in²   22 in²

3 in

4 in   23 in²

SOLUTION

PUZZLE 214

Find the solution on page 366.

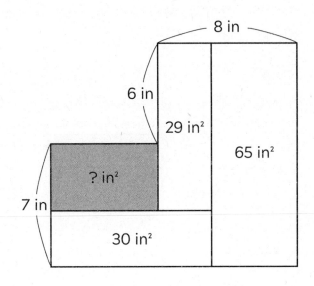

8 in

6 in

29 in²

65 in²

? in²

7 in

30 in²

SOLUTION

PUZZLE 215

Find the solution on page 367.

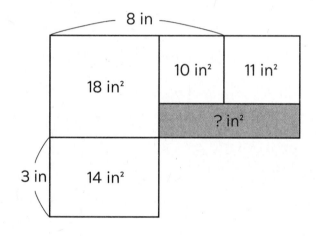

SOLUTION

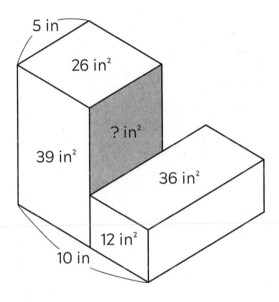

5 in

26 in²

? in²

39 in²

36 in²

12 in²

10 in

SOLUTION

PUZZLE 217

Find the solution on page 367.

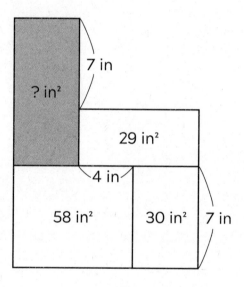

? in²

7 in

29 in²

4 in

58 in²   30 in²   7 in

SOLUTION

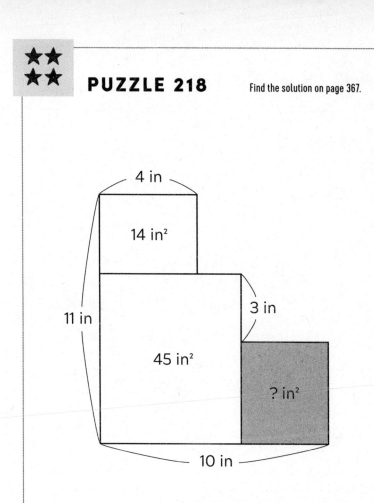

4 in

14 in²

11 in

3 in

45 in²

? in²

10 in

SOLUTION

PUZZLE 219

Find the solution on page 368.

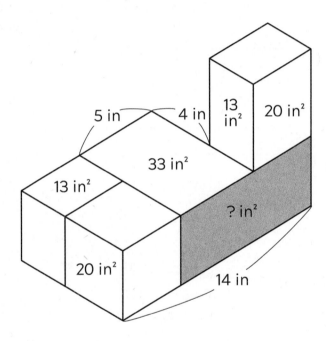

5 in   4 in   13 in²   20 in²

33 in²

13 in²

20 in²

? in²

14 in

SOLUTION

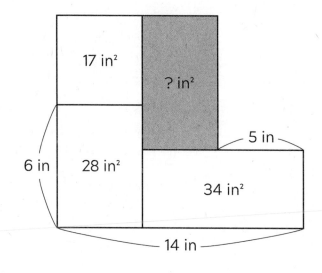

17 in²

? in²

5 in

6 in

28 in²

34 in²

14 in

SOLUTION

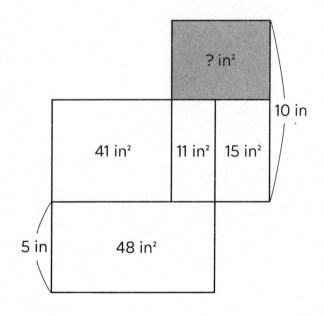

? in²

10 in

41 in²   11 in²   15 in²

5 in   48 in²

SOLUTION

PUZZLE 222

Find the solution on page 369.

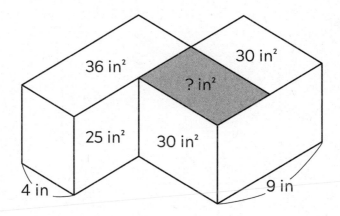

36 in²

30 in²

? in²

25 in²

30 in²

4 in

9 in

SOLUTION

PUZZLE 223

Find the solution on page 369.

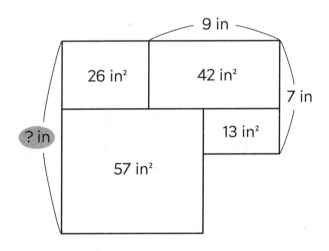

SOLUTION

PUZZLE 224

Find the solution on page 369.

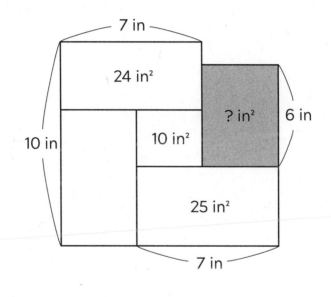

7 in

24 in²

? in²

6 in

10 in

10 in²

25 in²

7 in

SOLUTION

PUZZLE 225

Find the solution on page 369.

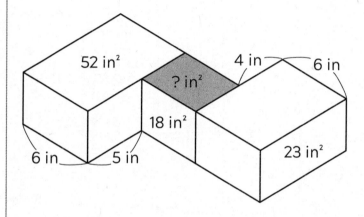

52 in²

4 in — 6 in

? in²

18 in²

6 in — 5 in

23 in²

SOLUTION

PUZZLE 226

Find the solution on page 370.

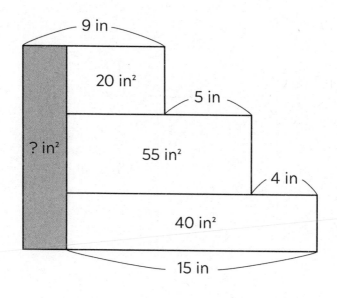

SOLUTION

PUZZLE 227

Find the solution on page 370.

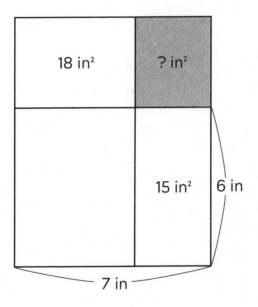

SOLUTION

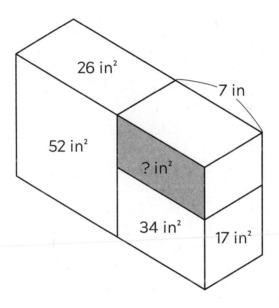

26 in²

7 in

52 in²

? in²

34 in²

17 in²

SOLUTION

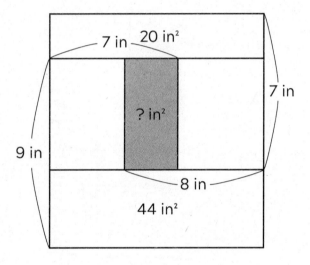

7 in   20 in²

7 in

? in²

9 in

8 in

44 in²

SOLUTION

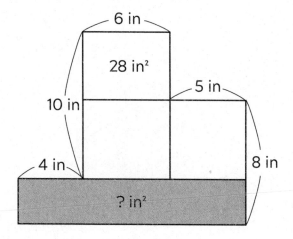

6 in

28 in²

5 in

10 in

4 in

8 in

? in²

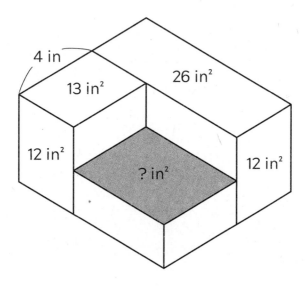

SOLUTION

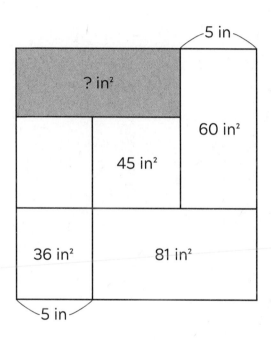

PUZZLE 233

Find the solution on page 371.

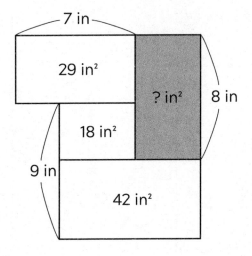

SOLUTION

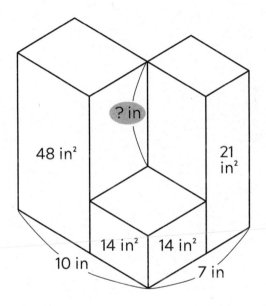

48 in²

21 in²

? in

14 in² 14 in²

10 in

7 in

SOLUTION

PUZZLE 235

Find the solution on page 372.

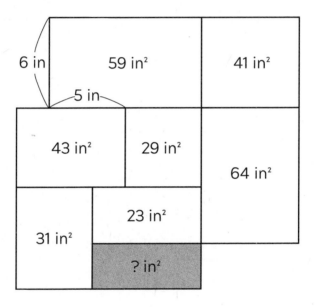

SOLUTION

PUZZLE 236

Find the solution on page 372.

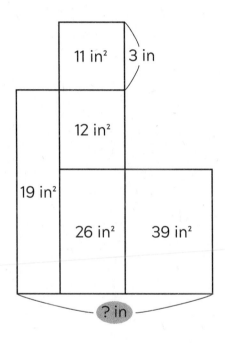

SOLUTION

PUZZLE 237

Find the solution on page 373.

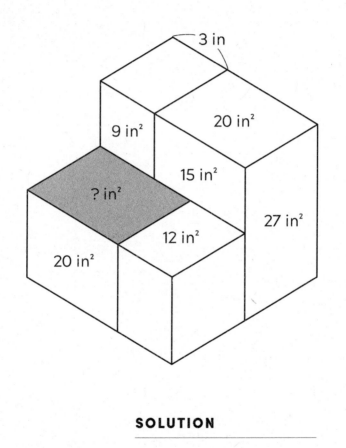

3 in

9 in²

20 in²

15 in²

? in²

27 in²

20 in²

12 in²

SOLUTION

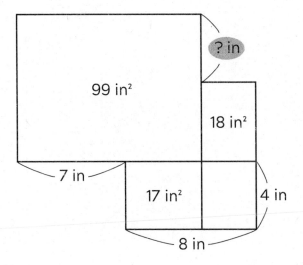

99 in²

? in

18 in²

7 in

17 in²

4 in

8 in

SOLUTION

PUZZLE 239

Find the solution on page 373.

9 in²		20 in²
	13 in²	
40 in²	? in²	40 in²
8 in²		

SOLUTION

PUZZLE 240

Find the solution on page 374.

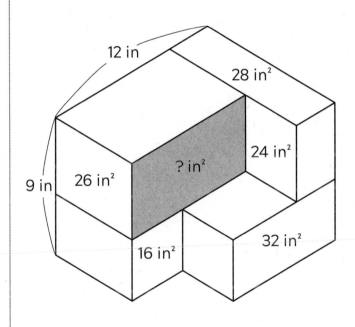

12 in

28 in²

24 in²

? in²

9 in   26 in²

16 in²   32 in²

SOLUTION

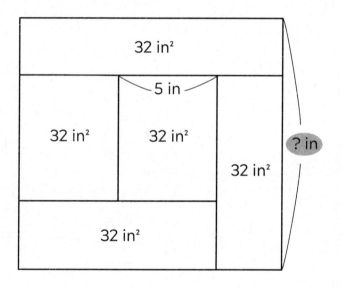

SOLUTION

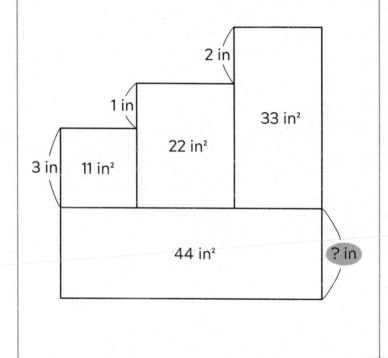

2 in

1 in

33 in²

22 in²

3 in   11 in²

44 in²   ? in

SOLUTION

PUZZLE 243

Find the solution on page 375.

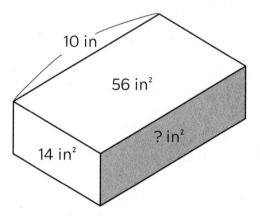

10 in

56 in²

14 in²

? in²

SOLUTION

PUZZLE 244

Find the solution on page 375.

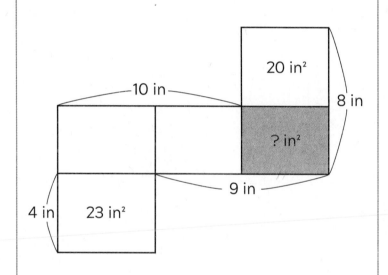

20 in²

10 in

8 in

? in²

9 in

4 in   23 in²

SOLUTION

PUZZLE 245

Find the solution on page 375.

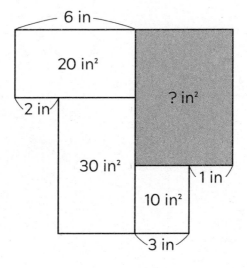

SOLUTION

PUZZLE 246

Find the solution on page 375.

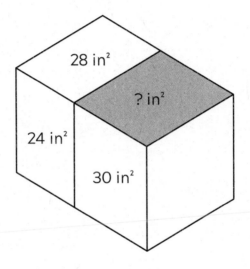

28 in²

? in²

24 in²

30 in²

SOLUTION

PUZZLE 247

Find the solution on page 376.

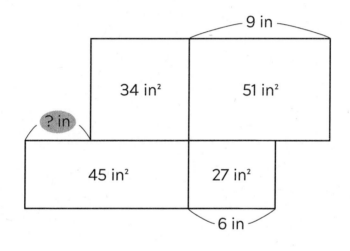

SOLUTION

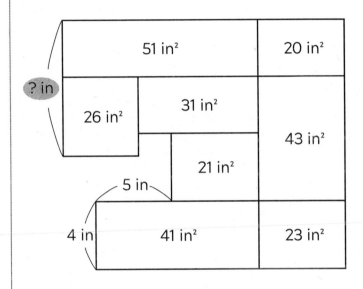

PUZZLE 249

Find the solution on page 376.

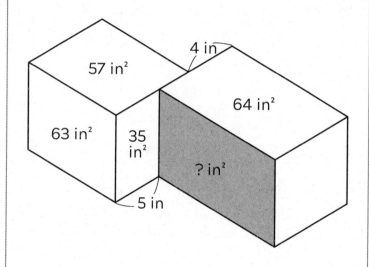

57 in²

4 in

64 in²

63 in²

35 in²

? in²

5 in

SOLUTION

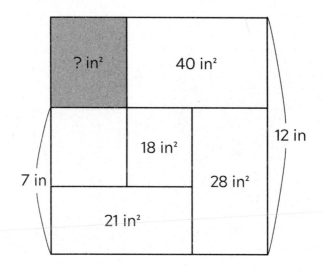

SOLUTION

_____

PUZZLE 251

Find the solution on page 377.

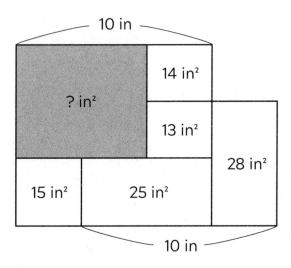

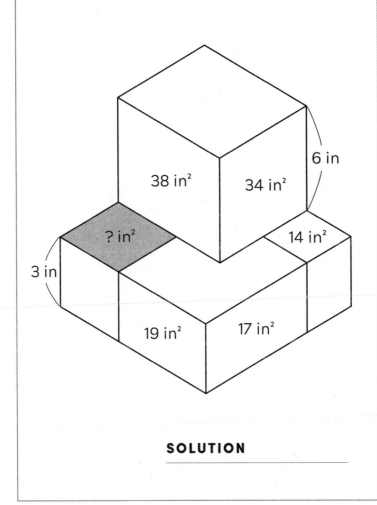

SOLUTION

PUZZLE 253

Find the solution on page 378.

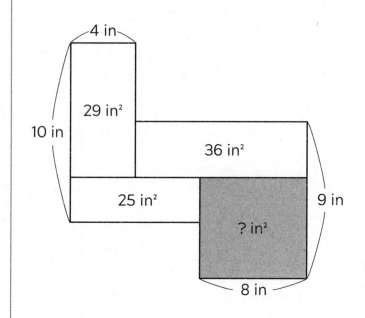

4 in

29 in²

10 in

36 in²

25 in²

9 in

? in²

8 in

SOLUTION

PUZZLE 254

Find the solution on page 378.

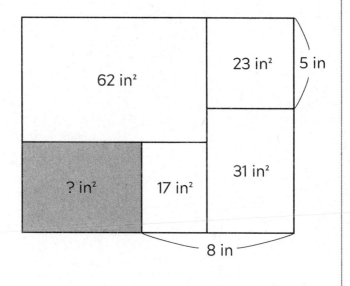

SOLUTION

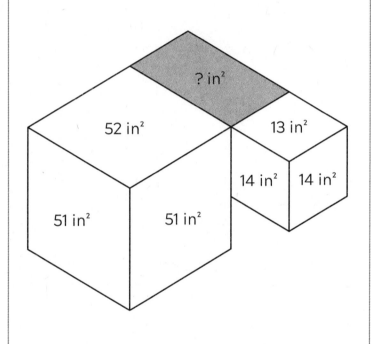

? in²

52 in²   13 in²

14 in²   14 in²

51 in²   51 in²

SOLUTION

PUZZLE 256

Find the solution on page 379.

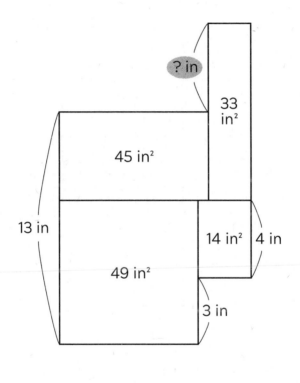

? in

33 in²

45 in²

13 in

14 in² 4 in

49 in²

3 in

SOLUTION

PUZZLE 257

Find the solution on page 379.

27 in²	27 in²

9 in²

? in²

9 in²

19 in²

37 in²

SOLUTION

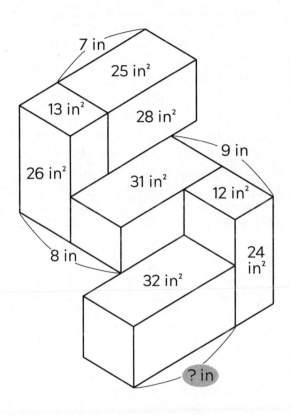

7 in

25 in²

13 in²

28 in²

26 in²

31 in²

9 in

12 in²

8 in

24 in²

32 in²

? in

SOLUTION

PUZZLE 259

Find the solution on page 380.

	66 in²	29 in²
? in²	41 in²	
		41 in²
72 in²		
		64 in²
	37 in²	

SOLUTION

PUZZLE 260

Find the solution on page 380.

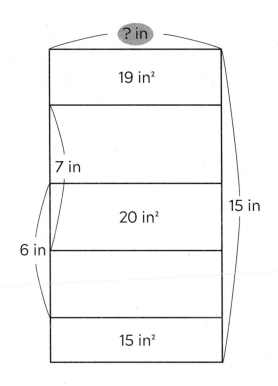

? in

19 in²

7 in

20 in²

15 in

6 in

15 in²

15 in²

SOLUTION

_____

PUZZLE 261

Find the solution on page 380.

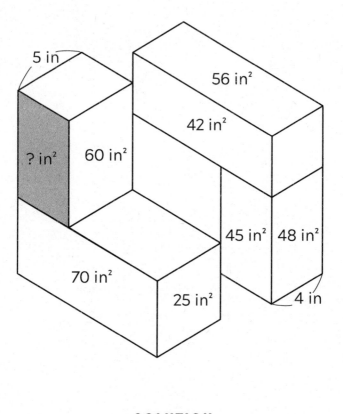

5 in

56 in²

42 in²

? in²   60 in²

45 in²   48 in²

70 in²

25 in²   4 in

SOLUTION

PUZZLE 262

Find the solution on page 381.

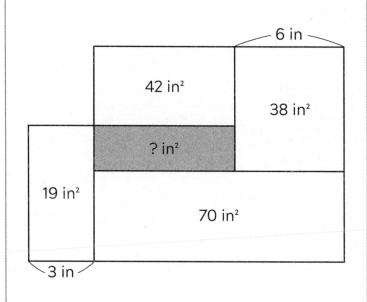

SOLUTION

PUZZLE 263

Find the solution on page 381.

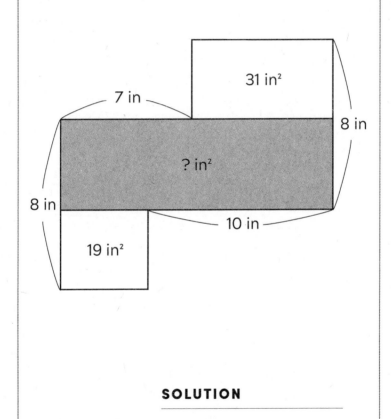

31 in²

7 in

8 in

? in²

8 in

10 in

19 in²

SOLUTION

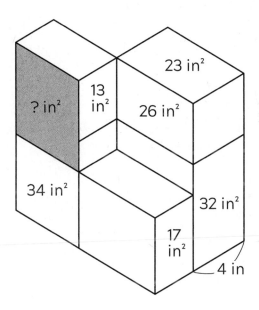

23 in²

13 in²

? in²

26 in²

34 in²

32 in²

17 in²

4 in

SOLUTION

PUZZLE 265

Find the solution on page 382.

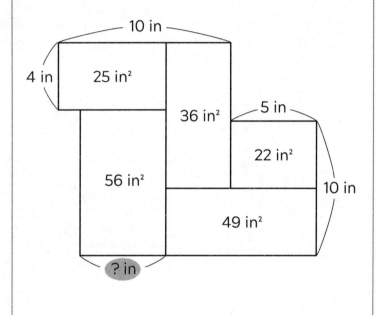

SOLUTION

PUZZLE 266

Find the solution on page 382.

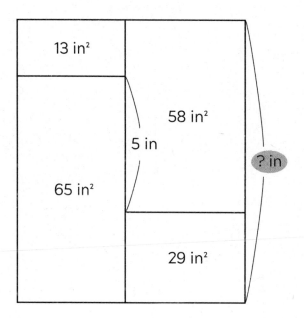

13 in²

58 in²

5 in

? in

65 in²

29 in²

SOLUTION

PUZZLE 267

Find the solution on page 382.

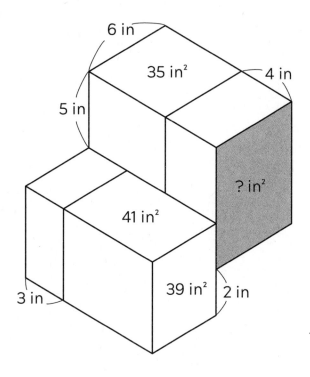

6 in

35 in²

4 in

5 in

? in²

41 in²

3 in

39 in²

2 in

SOLUTION

PUZZLE 268

Find the solution on page 383.

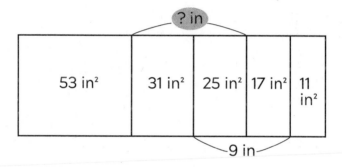

? in

| 53 in² | 31 in² | 25 in² | 17 in² | 11 in² |

9 in

SOLUTION

PUZZLE 269

Find the solution on page 383.

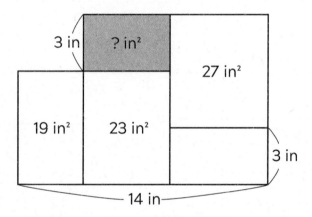

3 in

? in²

27 in²

19 in²

23 in²

3 in

14 in

SOLUTION

PUZZLE 270

Find the solution on page 383.

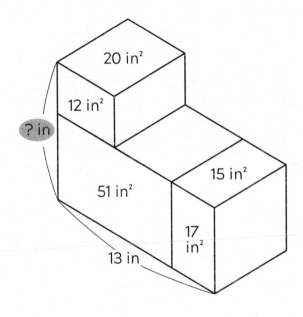

20 in²

12 in²

? in

51 in²

15 in²

17 in²

13 in

SOLUTION

PUZZLE 271

Find the solution on page 384.

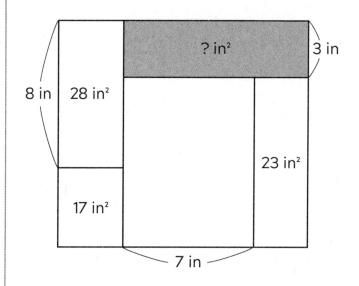

PUZZLE 272

Find the solution on page 384.

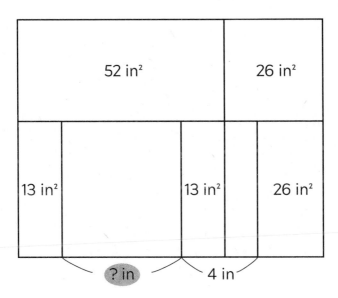

LEVEL 5 ◂ 279

PUZZLE 273

Find the solution on page 384.

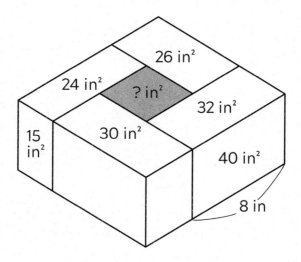

26 in²

24 in²

? in²

32 in²

15 in²

30 in²

40 in²

8 in

SOLUTION

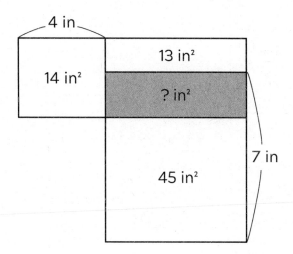

4 in

13 in²

14 in²

? in²

7 in

45 in²

SOLUTION

PUZZLE 275

Find the solution on page 385.

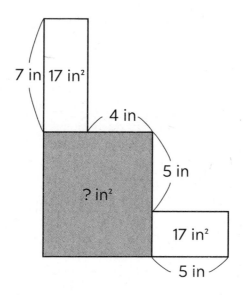

7 in | 17 in²

4 in

5 in

? in²

17 in²

5 in

SOLUTION

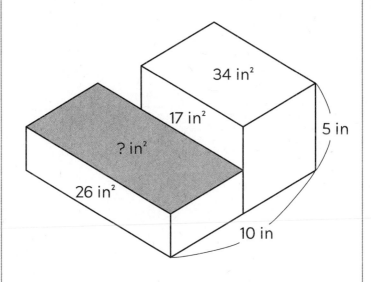

34 in²

17 in²

? in²

5 in

26 in²

10 in

SOLUTION

PUZZLE 277

Find the solution on page 386.

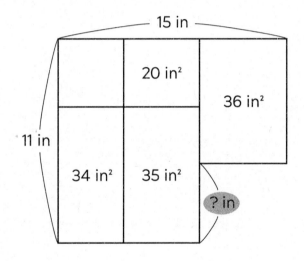

15 in

20 in²

36 in²

11 in

34 in²

35 in²

? in

SOLUTION

PUZZLE 278

Find the solution on page 386.

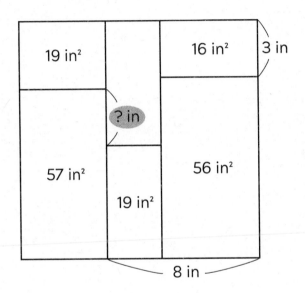

19 in²

16 in²

3 in

? in

57 in²

19 in²

56 in²

8 in

SOLUTION

PUZZLE 279

Find the solution on page 386.

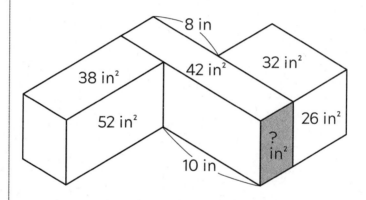

8 in

38 in²

42 in²

32 in²

52 in²

26 in²

? in²

10 in

SOLUTION

PUZZLE 280

Find the solution on page 387.

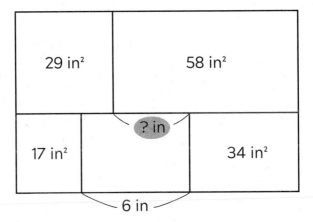

29 in²

58 in²

17 in²

? in

34 in²

6 in

SOLUTION

PUZZLE 281

Find the solution on page 387.

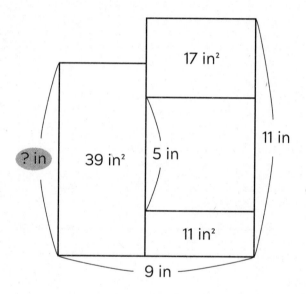

17 in²

? in

39 in²

5 in

11 in

11 in²

9 in

SOLUTION

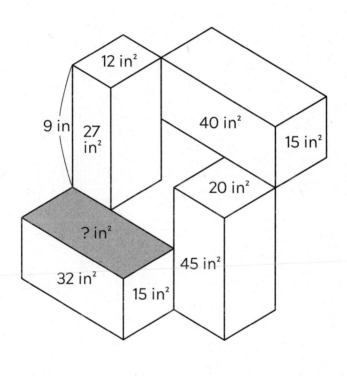

12 in²

9 in

27 in²

40 in²

15 in²

20 in²

? in²

45 in²

32 in²

15 in²

SOLUTION

Find the solution on page 388.

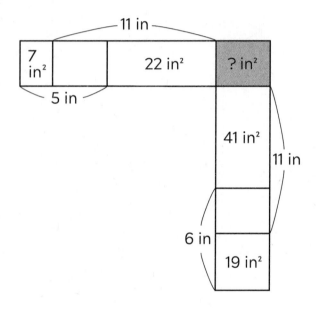

SOLUTION

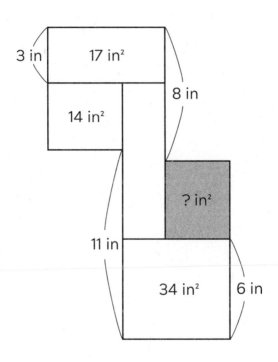

3 in | 17 in²

8 in

14 in²

? in²

11 in

34 in²   6 in

SOLUTION

PUZZLE 285

Find the solution on page 388.

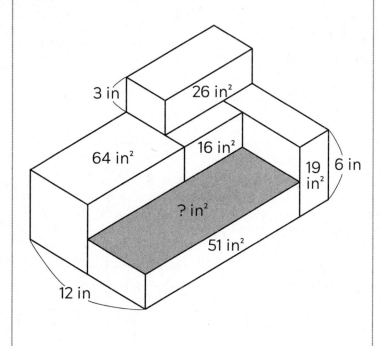

3 in

26 in²

64 in²

16 in²

19 in²

6 in

? in²

51 in²

12 in

SOLUTION

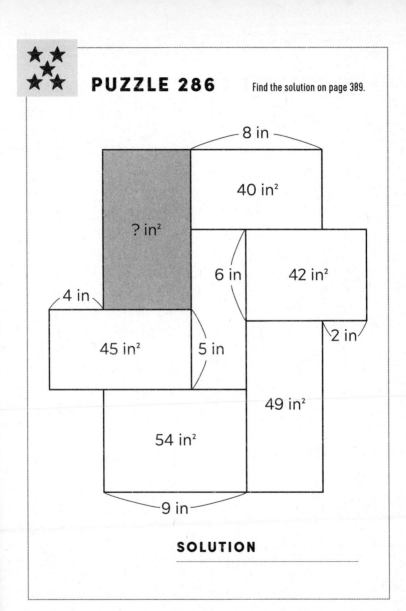

8 in

40 in²

? in²

6 in   42 in²

4 in

45 in²   5 in   2 in

49 in²

54 in²

9 in

SOLUTION

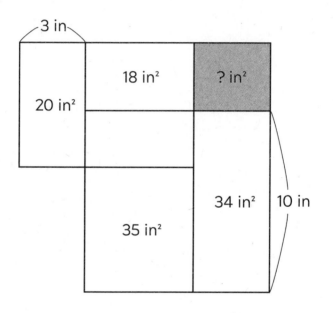

PUZZLE 288

Find the solution on page 390.

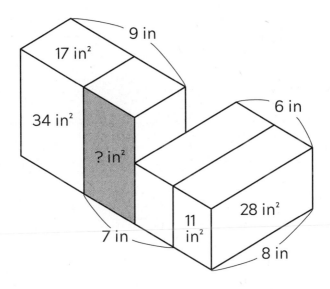

9 in

17 in²

34 in²

? in²

6 in

28 in²

11 in²

7 in

8 in

SOLUTION

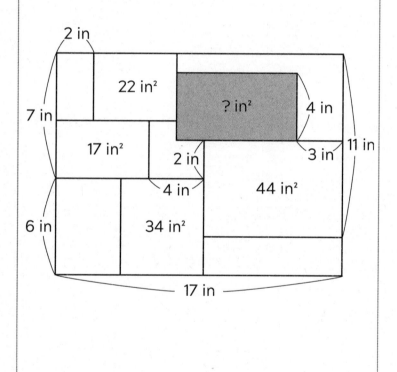

2 in

7 in

22 in²

? in²

4 in

6 in

17 in²

2 in

4 in

34 in²

44 in²

3 in

11 in

17 in

SOLUTION

PUZZLE 290

Find the solution on page 390.

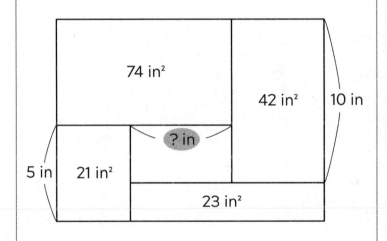

74 in²

42 in²

10 in

? in

5 in

21 in²

23 in²

SOLUTION

PUZZLE 291

Find the solution on page 391.

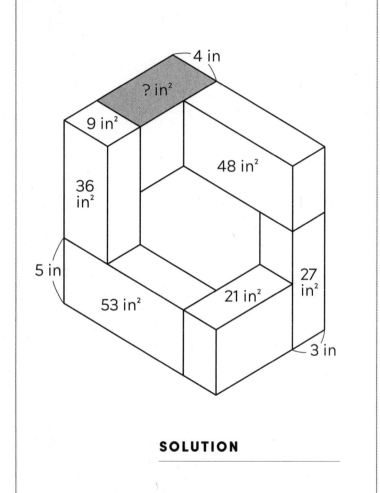

4 in

? in²

9 in²

48 in²

36 in²

5 in

27 in²

53 in²

21 in²

3 in

SOLUTION

PUZZLE 292

Find the solution on page 391.

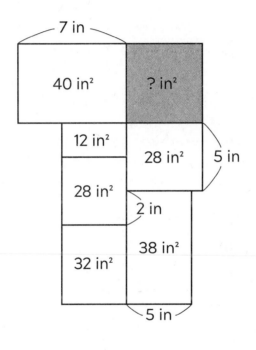

7 in

40 in²

? in²

12 in²

28 in²

28 in²

5 in

2 in

38 in²

32 in²

5 in

SOLUTION

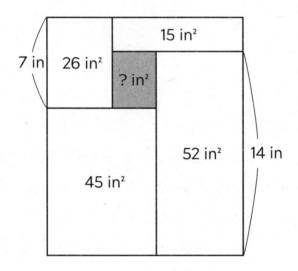

SOLUTION

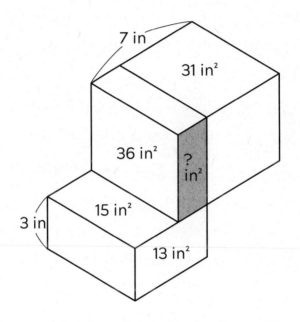

7 in

31 in²

36 in²

? in²

3 in

15 in²

13 in²

SOLUTION

_____

PUZZLE 295

Find the solution on page 392.

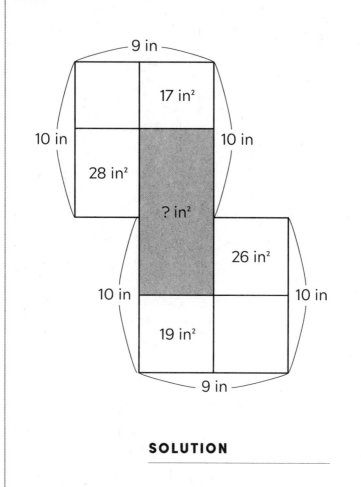

9 in

17 in²

10 in

10 in

28 in²

? in²

26 in²

10 in

10 in

19 in²

9 in

SOLUTION

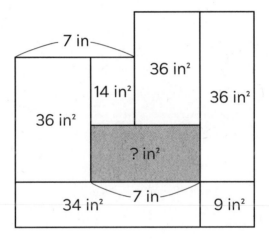

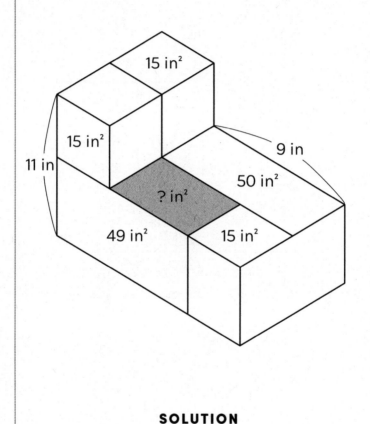

15 in²

15 in²

9 in

11 in

50 in²

? in²

49 in²

15 in²

SOLUTION

PUZZLE 298

Find the solution on page 394.

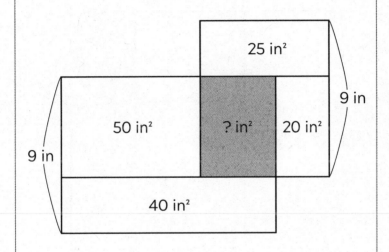

25 in²

9 in

50 in²   ? in²   20 in²

9 in

40 in²

SOLUTION

PUZZLE 299

Find the solution on page 395.

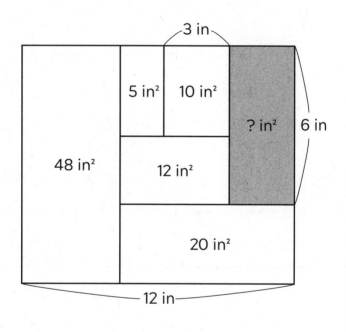

SOLUTION

_____

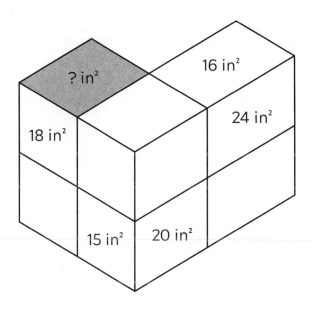

? in²

16 in²

24 in²

18 in²

15 in²   20 in²

SOLUTION

SOLUTIONS

PUZZLE 1

Ⓐ . . . 30 ÷ 5 = 6 in.
Length ⑦ is 42 ÷ Ⓐ = 7 in.

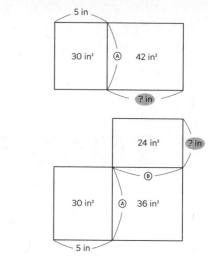

PUZZLE 2

Ⓐ . . . 30 ÷ 5 = 6 in.
Ⓑ . . . 36 ÷ 6 = 6 in.
Length ⑦ is 24 ÷ Ⓑ = 4 in.

PUZZLE 3

Ⓐ . . . 20 ÷ 5 = 4 in.
Ⓑ . . . 24 ÷ 4 = 6 in.
Area ⑦ is Ⓑ × 5 = 30 in.²

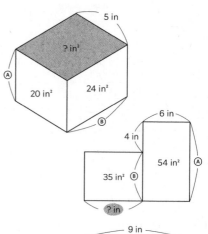

PUZZLE 4

Ⓐ . . . 54 ÷ 6 = 9 in.
Ⓑ . . . 9 − 4 = 5 in.
Length ⑦ is 35 ÷ Ⓑ = 7 in.

PUZZLE 5

Add the given areas: 24 + 30 = 54 in.²
Ⓐ . . . 54 ÷ 9 = 6 in.
Length ⑦ is 30 ÷ Ⓐ = 5 in.

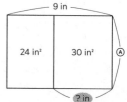

PUZZLE 6

Ⓐ . . . 56 ÷ 8 = 7 in.
Ⓑ . . . 42 ÷ 7 = 6 in.
Ⓒ . . . 30 ÷ 6 = 5 in.
Area ⑦ is Ⓐ × Ⓒ = 35 in.²

PUZZLE 7

Ⓐ . . . 13 − 6 = 7 in.
Ⓑ . . . 42 ÷ 7 = 6 in.
Ⓒ . . . 10 − 6 = 4 in.
Area ⑦ is Ⓒ × 6 = 24 in.²

PUZZLE 8

Ⓐ . . . 15 ÷ 3 = 5 in.
Ⓑ . . . 5 + 4 − 3 = 6 in.
Length ⑦ is 24 ÷ Ⓑ = 4 in.

PUZZLE 9

Ⓐ . . . 63 ÷ 9 = 7 in.
Ⓑ . . . 42 ÷ 7 = 6 in.
Ⓒ . . . 30 ÷ 6 = 5 in.
Ⓓ . . . 40 ÷ 5 = 8 in.
Area ⑦ is Ⓑ × Ⓓ = 48 in.²

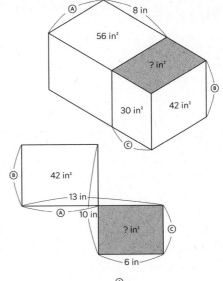

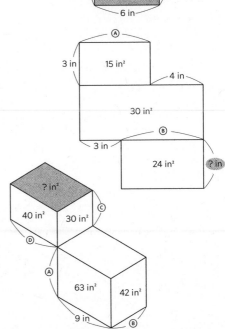

PUZZLE 10

Ⓐ ... 30 ÷ 6 = 5 in.
Ⓑ ... 25 ÷ 5 = 5 in.
Ⓒ ... 24 ÷ 6 = 4 in.
Area ⑦ is Ⓑ × Ⓒ = 20 in.²

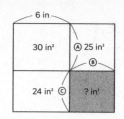

PUZZLE 11

Find the total area: 7 × 8 = 56 in.²
Area ⑦ is 56 − 24 − 18 = 14 in.²

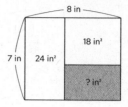

PUZZLE 12

Ⓐ ... 42 ÷ 7 = 6 in.
Ⓑ ... 36 ÷ 6 = 6 in.
Ⓒ ... 30 ÷ 6 = 5 in.
Ⓓ ... 40 ÷ 5 = 8 in.
Ⓔ ... 56 ÷ 8 = 7 in.
Area ⑦ is 7 × Ⓔ = 49 in.²

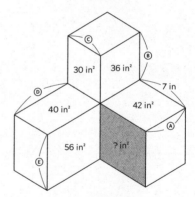

PUZZLE 13

Ⓐ ... 12 ÷ 3 = 4 in.
Ⓑ ... 11 − 4 = 7 in.
Ⓒ ... 49 ÷ 7 = 7 in.
Ⓓ ... 7 − 3 = 4 in.
Area ⑦ is Ⓐ × Ⓓ = 16 in.²

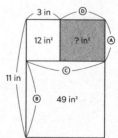

PUZZLE 14

Add the top two areas: 29 + 20 = 49 in.²
Ⓐ ... 49 ÷ 7 = 7 in.
Add the bottom two areas: 20 + 22 = 42 in.²
Length ⑦ is 42 ÷ Ⓐ = 6 in.

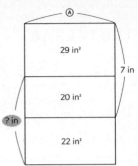

PUZZLE 15

Ⓐ ... 15 ÷ 3 = 5 in.
Ⓑ ... 20 ÷ 5 = 4 in.
Ⓒ ... 28 ÷ 4 = 7 in.
Ⓓ ... 63 ÷ 7 = 9 in.
Ⓔ ... 54 ÷ 9 = 6 in.
Ⓕ ... 24 ÷ 6 = 4 in.
Area ⑦ is Ⓓ × Ⓕ = 36 in.²

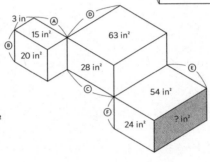

PUZZLE 16

Ⓐ ... 42 ÷ 7 = 6 in.
Ⓑ ... 48 ÷ 6 = 8 in.
Ⓒ ... 40 ÷ 8 = 5 in.
Length ⑦ is 45 ÷ Ⓒ = 9 in.

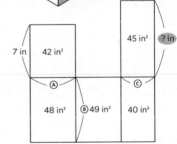

PUZZLE 17

Ⓐ ... 10 − 6 = 4 in.
Ⓑ ... 24 ÷ 4 = 6 in.
Ⓒ ... 6 + 5 = 11 in.
Area ⑦ is Ⓒ × 6 = 66 in.²

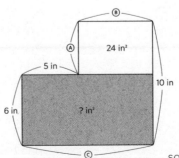

PUZZLE 18

Ⓐ ... 36 ÷ 4 = 9 in.
Ⓑ ... 45 ÷ 9 = 5 in.
Ⓒ ... 30 ÷ 5 = 6 in.
Ⓓ ... 36 ÷ 6 = 6 in.
Ⓔ ... 48 ÷ 6 = 8 in.
Ⓕ ... 40 ÷ 8 = 5 in.
Area ? is 4 × Ⓕ = 20 in.²

PUZZLE 19

Ⓐ ... 35 ÷ 5 = 7 in.
Ⓑ ... 28 ÷ 7 = 4 in.
Ⓒ ... 16 ÷ 4 = 4 in.
Area ? is Ⓒ × 5 = 20 in.²

PUZZLE 20

Ⓐ ... 36 ÷ 4 = 9 in.
Ⓑ ... 9 – 2 = 7 in.
Ⓒ ... 7 + 3 – 4 = 6 in.
Ⓓ ... 6 + 2 = 8 in.
Length ? is 24 ÷ Ⓓ = 3 in.

PUZZLE 21

Ⓐ ... 49 ÷ 7 = 7 in.
Ⓑ ... 56 ÷ 7 = 8 in.
Ⓒ ... 64 ÷ 8 = 8 in.
Ⓓ ... 72 ÷ 8 = 9 in.
Ⓔ ... 54 ÷ 9 = 6 in.
Area ? is 7 × Ⓔ = 42 in.²

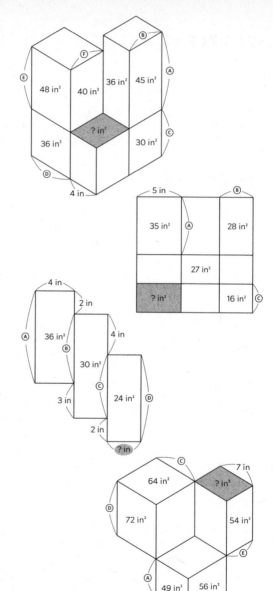

PUZZLE 22

Ⓐ ... 12 ÷ 3 = 4 in.
Ⓑ ... 32 ÷ 4 = 8 in.
Ⓒ ... 15 ÷ 3 = 5 in.
Ⓓ ... 20 ÷ 5 = 4 in.
Ⓔ ... 24 ÷ 4 = 6 in.
Area ⑦ is Ⓑ × Ⓔ = 48 in.²

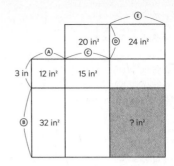

PUZZLE 23

Ⓐ ... 9 ÷ 3 = 3 in.
Ⓑ ... 7 − 3 = 4 in.
Ⓒ ... 20 ÷ 4 = 5 in.
Ⓓ ... 9 − 5 = 4 in.
Ⓔ ... 16 ÷ 4 = 4 in.
Ⓕ ... 8 − 4 = 4 in.
Area ⑦ is Ⓕ × 3 = 12 in.²

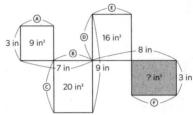

PUZZLE 24

Ⓐ ... 36 ÷ 6 = 6 in.
Ⓑ ... 30 ÷ 6 = 5 in.
Ⓒ ... 45 ÷ 5 = 9 in.
Ⓓ ... 36 ÷ 9 = 4 in.
Ⓔ ... 16 ÷ 4 = 4 in.
Ⓕ ... 12 ÷ 4 = 3 in.
Area ⑦ is Ⓒ × Ⓕ = 27 in.²

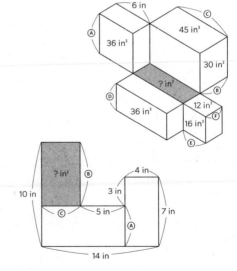

PUZZLE 25

Ⓐ ... 7 − 3 = 4 in.
Ⓑ ... 10 − 4 = 6 in.
Ⓒ ... 14 − 4 − 5 = 5 in.
Area ⑦ is Ⓑ × Ⓒ = 30 in.²

PUZZLE 26

Draw areas Ⓐ and Ⓑ.

Ⓐ . . . (8 × 6) – 21 = 27 in.²

Ⓑ . . . (7 × 4) – 10 = 18 in.²

Area ⁇ is Ⓐ + Ⓑ = 45 in.²

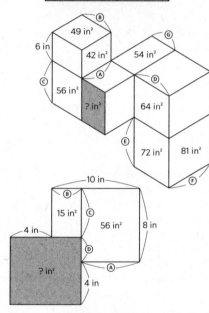

PUZZLE 27

Ⓐ . . . 42 ÷ 6 = 7 in.

Ⓑ . . . 49 ÷ 7 = 7 in.

Ⓒ . . . 56 ÷ 7 = 8 in.

Ⓓ . . . 64 ÷ 8 = 8 in.

Ⓔ . . . 72 ÷ 8 = 9 in.

Ⓕ . . . 81 ÷ 9 = 9 in.

Ⓖ . . . 54 ÷ 9 = 6 in.

Area ⁇ is Ⓒ × Ⓖ = 48 in.²

PUZZLE 28

Ⓐ . . . 56 ÷ 8 = 7 in.

Ⓑ . . . 10 – 7 = 3 in.

Ⓒ . . . 15 ÷ 3 = 5 in.

Ⓓ . . . 8 – 5 = 3 in.

Area ⁇ is (Ⓑ + 4) × (Ⓓ + 4) = 49 in.²

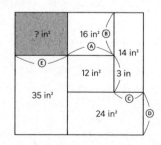

PUZZLE 29

Ⓐ . . . 12 ÷ 3 = 4 in.

Ⓑ . . . 16 ÷ 4 = 4 in.

Ⓒ . . . 14 ÷ (4 + 3) = 2 in.

Ⓓ . . . 24 ÷ (4 + 2) = 4 in.

Ⓔ . . . 35 ÷ (3 + 4) = 5 in.

Area ⁇ is Ⓑ × Ⓔ = 20 in.²

PUZZLE 30

Ⓐ ... 24 ÷ 6 = 4 in.
Ⓑ ... 28 ÷ 4 = 7 in.
Ⓒ ... 56 ÷ 7 = 8 in.
Ⓓ ... 30 ÷ 6 = 5 in.
Ⓔ ... 45 ÷ 5 = 9 in.
Area ⓐ is Ⓒ × Ⓔ = 72 in.²

PUZZLE 31

Ⓐ ... 32 ÷ 8 = 4 in.
Ⓑ ... 54 ÷ (5 + 4) = 6 in.
Ⓒ ... 42 ÷ 6 = 7 in.
Ⓓ ... 7 − 3 = 4 in.
Length ⓐ is 40 ÷ Ⓓ = 10 in.

PUZZLE 32

Ⓐ ... 24 ÷ 6 = 4 in.
Ⓑ ... 20 ÷ 4 = 5 in.
Ⓒ ... 25 ÷ 5 = 5 in.
Ⓓ ... 36 ÷ 6 = 6 in.
Ⓔ ... 42 ÷ 6 = 7 in.
Area ⓐ is Ⓒ × Ⓔ = 35 in.²

PUZZLE 33

Ⓐ ... 42 ÷ 6 = 7 in.
Ⓑ ... 49 ÷ 7 = 7 in.
Ⓒ ... 56 ÷ 7 = 8 in.
Ⓓ ... 8 − 6 = 2 in.
Ⓔ ... 2 + 7 = 9 in.
Ⓕ ... 45 ÷ 9 = 5 in.
Area ⓐ is Ⓐ × Ⓕ = 35 in.²

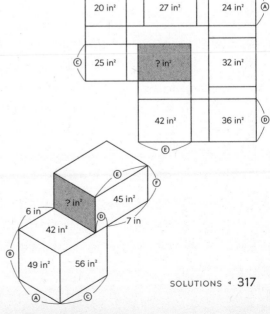

PUZZLE 34

Ⓐ ... 35 ÷ 7 = 5 in.
Ⓑ ... 11 − 5 = 6 in.
Ⓒ ... 24 ÷ 6 = 4 in.
Ⓓ ... 7 − 4 = 3 in.
Ⓔ ... 6 − 3 = 3 in.
Area ⑦ is (2 + Ⓓ) × Ⓔ = 15 in.²

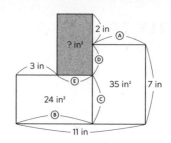

PUZZLE 35

Ⓐ ... 9 − 6 = 3 in.
Ⓑ ... 10 − 3 − 3 = 4 in.
Ⓒ ... 11 − 3 = 8 in.
Ⓓ ... 8 − 4 = 4 in.
Ⓔ ... 7 − 4 = 3 in.
Area ⑦ is Ⓑ × Ⓔ = 12 in.²

PUZZLE 36

Ⓐ ... 30 ÷ 6 = 5 in.
Ⓑ ... 35 ÷ 5 = 7 in.
Ⓒ ... 10 − 6 = 4 in.
Ⓓ ... 24 ÷ 4 = 6 in.
Area ⑦ is Ⓑ × Ⓓ = 42 in.²

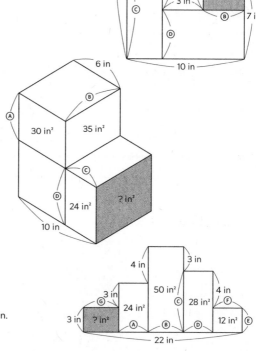

PUZZLE 37

Ⓐ ... 24 ÷ (3 + 3) = 4 in.
Ⓑ ... 50 ÷ (4 + 3 + 3) = 5 in.
Ⓒ ... (4 + 3 + 3) − 3 = 7 in.
Ⓓ ... 28 ÷ 7 = 4 in.
Ⓔ ... 7 − 4 = 3 in.
Ⓕ ... 12 ÷ 3 = 4 in.
Ⓖ ... 22 − (4 + 5 + 4 + 4) = 5 in.
Area ⑦ is Ⓖ × 3 = 15 in.²

PUZZLE 38

Ⓐ ... 9 − 5 = 4 in.
Ⓑ ... 28 ÷ 4 = 7 in.
Ⓒ ... 35 ÷ 5 = 7 in.
Ⓓ ... (7 + 5) − 7 = 5 in.
Area ? is Ⓓ × 5 = 25 in.²

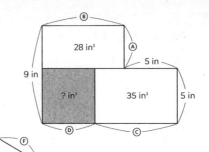

PUZZLE 39

Ⓐ ... 20 ÷ 5 = 4 in.
Ⓑ ... 24 ÷ 4 = 6 in.
Ⓒ ... 54 ÷ 6 = 9 in.
Ⓓ ... 9 − 5 = 4 in.
Ⓔ ... 4 + 3 = 7 in.
Ⓕ ... 35 ÷ 7 = 5 in.
Ⓖ ... 30 ÷ 5 = 6 in.
Area ? is 3 × Ⓖ = 18 in.²

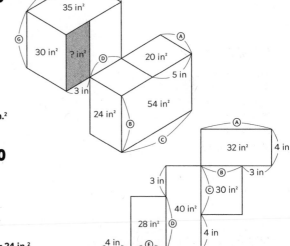

PUZZLE 40

Ⓐ ... 32 ÷ 4 = 8 in.
Ⓑ ... 8 − 3 = 5 in.
Ⓒ ... 30 ÷ 5 = 6 in.
Ⓓ ... (6 + 4) − 3 = 7 in.
Ⓔ ... 28 ÷ 7 = 4 in.
Area ? is 3 × (4 + Ⓔ) = 24 in.²

PUZZLE 41

Ⓐ ... 8 − 3 = 5 in.
Ⓑ ... 25 ÷ 5 = 5 in.
Ⓒ ... 42 ÷ 6 = 7 in.
Ⓓ ... 7 − (5 − 3) = 5 in.
Ⓔ ... 5 + 4 = 9 in.
Area ? is Ⓔ × 8 = 72 in.²

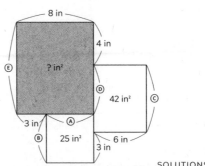

PUZZLE 42

(A) ... 63 ÷ 7 = 9 in.
(B) ... 9 − 5 + 4 = 8 in.
(C) ... 64 ÷ 8 = 8 in.
(D) ... 56 ÷ 8 = 7 in.
(E) ... 6 + 7 − 3 = 10 in.
Area (?) is 7 × (E) = 70 in.²

PUZZLE 43

(A) ... 9 ÷ 3 = 3 in.
(B) ... 28 ÷ (3 + 4) = 4 in.
(C) ... 56 ÷ (4 + 3) = 8 in.
(D) ... 12 ÷ 4 = 3 in.
(E) ... 8 − 4 = 4 in.
Area (?) is (E) × (3 + (D)) = 24 in.²

PUZZLE 44

(A) ... 42 ÷ 6 = 7 in.
(B) ... (7 − 3) + 4 = 8 in.
(C) ... 72 ÷ 8 = 9 in.
(D) ... (9 − 4) + 5 = 10 in.
(E) ... 60 ÷ 10 = 6 in.
(F) ... (6 − 3) + 4 = 7 in.
Length (?) is 35 ÷ (F) = 5 in.

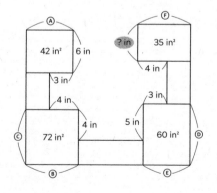

PUZZLE 45

Ⓐ . . . 24 ÷ 3 = 8 in.
Ⓑ . . . 32 ÷ 8 = 4 in.
Ⓒ . . . 28 ÷ 4 = 7 in.
Ⓓ . . . 7 – 3 = 4 in.
Ⓔ . . . 16 ÷ 4 = 4 in.
Ⓕ . . . 40 ÷ 8 = 5 in.
Ⓖ . . . 5 + 4 = 9 in.
Ⓗ . . . 18 ÷ 9 = 2 in.
Area ⑦ is Ⓐ × Ⓗ = 16 in.²

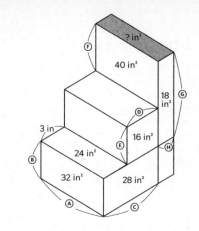

PUZZLE 46

Ⓐ . . . 48 ÷ 6 = 8 in.
Ⓑ . . . 14 – 8 = 6 in.
Ⓒ . . . 24 ÷ 6 = 4 in.
Ⓓ . . . (6 + 5) – 4 = 7 in.
Ⓔ . . . 25 ÷ 5 = 5 in.
Area ⑦ is Ⓓ × (9 – Ⓔ) = 28 in.²

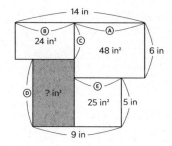

PUZZLE 47

Ⓐ . . . 3 + 3 = 6 in.
Ⓑ . . . 24 ÷ 6 = 4 in.
Ⓒ . . . 4 + 3 = 7 in.
Ⓓ . . . 14 ÷ 7 = 2 in.
Ⓔ . . . 30 ÷ (2 + 3) = 6 in.
Ⓕ . . . 6 – 3 = 3 in.
Ⓖ . . . 24 ÷ 3 = 8 in.
Ⓗ . . . 7 + 6 – 3 = 10 in.
Area ⑦ is Ⓗ × (Ⓖ – 3) = 50 in.²

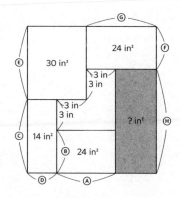

PUZZLE 48

Ⓐ ... 12 ÷ 4 = 3 in.
Ⓑ ... 9 ÷ 3 = 3 in.
Ⓒ ... 3 + 3 = 6 in.
Ⓓ ... 18 ÷ 6 = 3 in.
Ⓔ ... 21 ÷ 3 = 7 in.
Ⓕ ... 7 + 2 = 9 in.
Ⓖ ... 27 ÷ 9 = 3 in.
Ⓗ ... 4 + 6 = 10 in.
Area ⑦ is Ⓖ × Ⓗ = 30 in.²

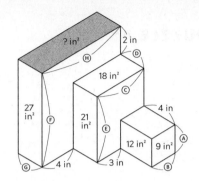

PUZZLE 49

Add the top two areas: 28 + 8 = 36 in.²
Ⓐ ... 36 ÷ 9 = 4 in.
Ⓑ ... 28 ÷ 4 = 7 in.
Add the middle two areas: 25 + 10 = 35 in.²
Ⓒ ... 35 ÷ 7 = 5 in.
Ⓓ ... 25 ÷ 5 = 5 in.
Ⓔ ... 20 ÷ 5 = 4 in.
Length ⑦ is Ⓐ + Ⓒ + Ⓔ = 13 in.

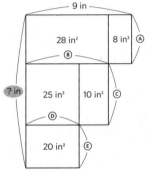

PUZZLE 50

Ⓐ ... 8 – 3 = 5 in.
Ⓑ ... 35 ÷ 5 = 7 in.
Ⓒ ... 21 ÷ 7 = 3 in.
Ⓓ ... 56 ÷ (3 + 5) = 7 in.
Ⓔ ... 7 – 3 = 4 in.
Ⓕ ... 28 ÷ 4 = 7 in.
Ⓖ ... 42 ÷ 7 = 6 in.
Area ⑦ is (Ⓖ + 3) × 8 = 72 in.²

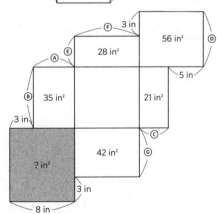

PUZZLE 51

Ⓐ ... 15 ÷ 5 = 3 in.
Ⓑ ... 9 ÷ 3 = 3 in.
Ⓒ ... 48 ÷ (3 + 5) = 6 in.
Ⓓ ... 3 + 6 = 9 in.
Ⓔ ... 45 ÷ 9 = 5 in.
Ⓕ ... 20 ÷ 5 = 4 in.
Ⓖ ... 24 ÷ 4 = 6 in.
Ⓗ ... 30 ÷ 6 = 5 in.
Area ⓐ is Ⓕ × (Ⓓ − Ⓗ) = 16 in.²

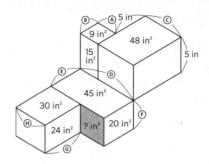

PUZZLE 52

Ⓐ ... 30 ÷ 5 = 6 in.
Ⓑ ... 16 ÷ 4 = 4 in.
Ⓒ ... 20 ÷ 5 = 4 in.
Lengths Ⓐ + Ⓑ span the same
  total height as Ⓒ + Ⓓ + the
  given height of 3.
Ⓓ ... 6 + 4 − 3 − 4 = 3 in.
Ⓔ ... 28 ÷ 4 = 7 in.
Ⓕ ... 15 ÷ 3 = 5 in.
Ⓖ ... 25 ÷ 5 = 5 in.
Lengths Ⓔ + Ⓕ span the same total
  width as Ⓖ + Ⓗ + the given width of 4.
Ⓗ ... 7 + 5 − 4 − 5 = 3 in.
Area ⓐ is Ⓓ × Ⓗ = 9 in.²

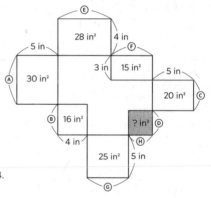

PUZZLE 53

Ⓐ ... (15 + 25) ÷ 8 = 5 in.
Ⓑ ... 25 ÷ 5 = 5 in.
Ⓒ ... 30 ÷ 5 = 6 in.
Ⓓ ... (42 + 24) ÷ (5 + 6) = 6 in.
Ⓔ ... 24 ÷ 6 = 4 in.
Ⓕ ... 16 ÷ 4 = 4 in.
Length ⓐ is Ⓓ + Ⓕ = 10 in.

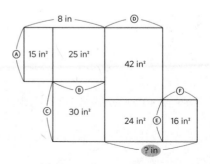

PUZZLE 54

Ⓐ . . . 45 ÷ 5 = 9 in.
Ⓑ . . . 9 − 6 = 3 in.
Ⓒ . . . (3 + 4) − 2 = 5 in.
Ⓓ . . . 5 + 3 = 8 in.
Ⓔ . . . 40 ÷ 8 = 5 in.
Ⓕ . . . 30 ÷ 5 = 6 in.
Ⓖ . . . 24 ÷ 6 = 4 in.
Area ? is (Ⓑ + 4) × Ⓖ = 28 in.²

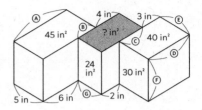

PUZZLE 55

Ⓐ . . . 9 ÷ 3 = 3 in.
Ⓑ . . . 35 ÷ (4 + 3) = 5 in.
Ⓒ . . . 10 − 5 = 5 in.
Ⓓ . . . 20 ÷ 5 = 4 in.
Ⓔ . . . 28 ÷ 4 = 7 in.
Ⓕ . . . 7 − 3 = 4 in.
Ⓖ . . . (36 ÷ 4) − 5 = 4 in.
Ⓗ . . . 16 ÷ 4 = 4 in.
Ⓘ . . . 11 − 4 − 4 = 3 in.
Area ? is Ⓘ × 3 = 9 in.²

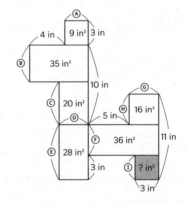

PUZZLE 56

Find the total area: 27 + 14 + 21 + 16 + 12 = 90 in.²
Ⓐ . . . 90 ÷ 10 = 9 in.
Ⓑ . . . 27 ÷ 9 = 3 in.
Ⓒ . . . 10 − 3 = 7 in.
Ⓓ . . . 14 ÷ 7 = 2 in.
Ⓔ . . . 21 ÷ (9 − 2) = 3 in.
Ⓕ . . . 7 − 3 = 4 in.
Length ? is 12 ÷ Ⓕ = 3 in.

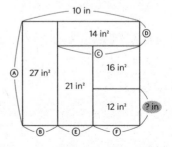

PUZZLE 57

(A) . . . 24 ÷ 6 = 4 in.
(B) . . . 20 ÷ 4 = 5 in.
(C) . . . 5 + 5 − 3 = 7 in.
(D) . . . 49 ÷ 7 = 7 in.
(E) . . . 42 ÷ 7 = 6 in.
(F) . . . 6 + 4 − 2 = 8 in.
(G) . . . 32 ÷ 8 = 4 in.
(H) . . . 28 ÷ 4 = 7 in.
Area ⑦ is (F) × (H) = 56 in.²

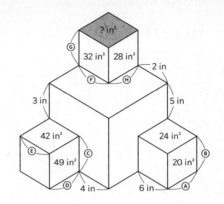

PUZZLE 58

(A) . . . 27 ÷ 9 = 3 in.
(B) . . . 9 − 6 = 3 in.
(C) . . . 9 ÷ 3 = 3 in.
(D) . . . 10 − 3 − 3 = 4 in.
(E) . . . 24 ÷ 4 = 6 in.
(F) . . . 9 − 6 = 3 in.
(G) . . . 30 ÷ 3 = 10 in.
Area ⑦ is ((E) − (B)) × ((G) − (D)) = 18 in.²

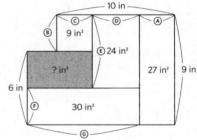

PUZZLE 59

(A) . . . 9 ÷ 3 = 3 in.
(B) . . . 8 − 3 = 5 in.
(C) . . . 25 ÷ 5 = 5 in.
(D) . . . 24 ÷ (9 − 5) = 6 in.
(E) . . . 10 − (36 ÷ 6) = 4 in.
(F) . . . 7 − (12 ÷ 4) = 4 in.
(G) . . . 8 − (20 ÷ 4) = 3 in.
(H) . . . 15 ÷ 3 = 5 in.
(H) + the given length of 7 span
the same total width as (C) + (I)
+ the given length of 3.
(I) . . . 5 + 7 − 5 − 3 = 4 in.
Similarly, (J) . . . 3 + 6 + 5 − 10 = 4 in.
Area ⑦ is (I) × (J) = 16 in.²

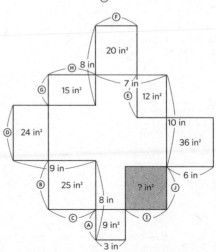

PUZZLE 60

Ⓐ ... 12 ÷ 6 = 2 in.
Ⓑ ... 4 + 5 − 2 = 7 in.
Ⓒ ... 28 ÷ 7 = 4 in.
Ⓓ ... 32 ÷ 4 = 8 in.
Ⓔ ... 15 ÷ 5 = 3 in.
Ⓕ ... 8 + 3 − 6 = 5 in.
Area ⑦ is 4 × Ⓕ = 20 in.²

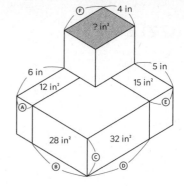

PUZZLE 61

Ⓐ ... 10 − 5 = 5 in.
Draw areas Ⓑ and Ⓒ.
The areas of Ⓑ and 24 are the same width and height: Ⓑ = 24 in.²
Ⓒ ... 49 − 24 = 25 in.²
Length ⑦ is Ⓒ ÷ 5 = 5 in.

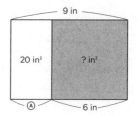

PUZZLE 62

Ⓐ ... 9 − 6 = 3 in.
The two areas are the same height, and 3 is half of 6.
Therefore 20 must be half of ⑦.
Area ⑦ is 40 in.²

PUZZLE 63

The two given areas share a side, and 26 is half of 52.
Therefore Ⓐ must be half of 8: Ⓐ = 4 in.
Area ⑦ is 8 × Ⓐ = 32 in.²

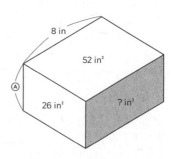

PUZZLE 64

The top two areas are the same height, and 20 is double 10.

Therefore Ⓐ must be double Ⓑ.

Similarly, Ⓒ must be double 15: Ⓒ = 30 in.²

Ⓓ . . . 50 − 30 = 20 in.²

Length ⑦ is Ⓓ ÷ 4 = 5 in.

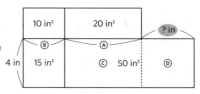

PUZZLE 65

Ⓐ . . . 7 − 3 = 4 in.

The given areas of 25 are the same height, so they must be the same width: Ⓑ = 4 in.

Length ⑦ is 8 − Ⓑ = 4 in.

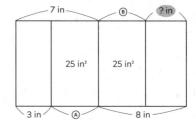

PUZZLE 66

The two areas on top share a side, and 36 is double 18.

Therefore Ⓐ must be double Ⓑ.

Similarly, ⑦ must be double 21.

Area ⑦ is 42 in.²

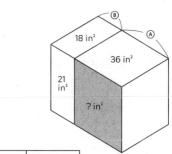

PUZZLE 67

The top two areas are the same height, and 38 is double 19.

Therefore Ⓐ must be double Ⓑ.

Similarly, Ⓒ must be double 21: Ⓒ = 42 in.²

This is triple 14, so Ⓓ must be triple Ⓔ.

Therefore 30 must be triple ⑦.

Area ⑦ is 10 in.²

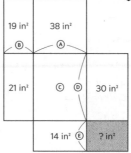

PUZZLE 68

Ⓐ ... 8 – 4 = 4 in.

Ⓑ and Ⓒ are the same height and width, so they must be equal.

Ⓒ ... 38 ÷ 2 = 19 in.²

Length ⑦ is (19 + 17) ÷ 4 = 9 in.

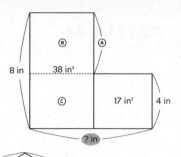

PUZZLE 69

Ⓐ ... (35 + 21) ÷ 8 = 7 in.

Ⓑ ... (27 + 36) ÷ 7 = 9 in.

Area ⑦ is (8 × Ⓑ) – 34 = 38 in.²

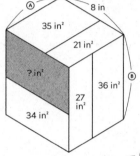

PUZZLE 70

Ⓐ ... 45 ÷ 5 = 9 in.

Ⓑ ... 13 – 9 = 4 in.

Ⓒ ... 28 ÷ 4 = 7 in.

Area ⑦ is ((Ⓐ × Ⓒ) – 32 = 31 in.²

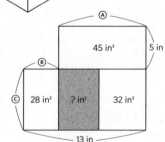

PUZZLE 71

Ⓐ ... 5 × 4 = 20 in.²

Ⓑ ... 10 – 5 = 5 in.

The areas of Ⓒ and 29 are the same height and width: Ⓒ = 29 in.²

Area ⑦ is Ⓐ + Ⓒ = 49 in.²

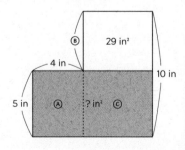

PUZZLE 72

Add the two areas on top: 30 + 42 = 72 in.^{2}

Ⓐ . . . 72 ÷ 12 = 6 in.

Ⓑ . . . 30 ÷ 6 = 5 in.

Ⓒ . . . 40 ÷ 5 = 8 in.

Area ⦿ is (Ⓐ × Ⓒ) − 14 = 34 in.^{2}

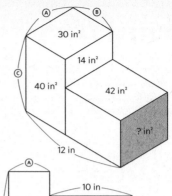

PUZZLE 73

Ⓐ . . . (45 + 20) ÷ 13 = 5 in.

Ⓑ . . . 9 − 5 = 4 in.

Ⓒ . . . 10 − 4 = 6 in.

Area ⦿ is (14 × Ⓒ) − 25 = 59 in.^{2}

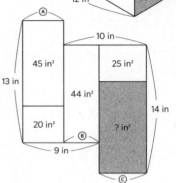

PUZZLE 74

Since 34 is double 17, Ⓐ must be double 13: Ⓐ = 26 in.^{2}

Find the total area: 13 + 26 + 17 + 34 = 90 in.^{2}

Length ⦿ is 90 ÷ 10 = 9 in.

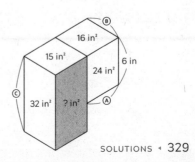

PUZZLE 75

Ⓐ . . . 24 ÷ 6 = 4 in.

Ⓑ . . . 16 ÷ 4 = 4 in.

Ⓒ . . . 32 ÷ 4 = 8 in.

The areas of 15 and ⦿ share a side, and their other sides are Ⓑ are Ⓒ.

Ⓒ is double Ⓑ, therefore ⦿ must be double 15.

Area ⦿ is 30 in.^{2}

PUZZLE 76

Ⓐ . . . 42 ÷ 7 = 6 in.
Add the areas on left: 21 + 42 = 63 in.²
Add the areas on right: 48 + 15 = 63 in.²
These are equal, so Ⓑ must equal Ⓐ: Ⓑ = 6 in.
Length ⑦ is 48 ÷ Ⓑ = 8 in.

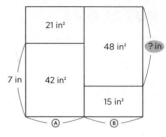

PUZZLE 77

Ⓐ . . . 16 ÷ 4 = 4 in.
Ⓑ . . . 20 ÷ 4 = 5 in.
Ⓒ . . . 25 ÷ 5 = 5 in.
Ⓓ . . . 30 ÷ 5 = 6 in.
Ⓔ . . . 24 ÷ 4 = 6 in.
Ⓕ . . . 36 ÷ 6 = 6 in.
This equals Ⓓ, so the areas of ⑦ and
  45 are the same height and width.
Area ⑦ is 45 in.²

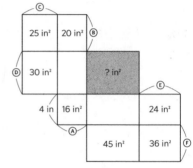

PUZZLE 78

Ⓐ . . . (24 + 25) ÷ 7 = 7 in.
Length ⑦ is (27 + 29) ÷ Ⓐ = 8 in.

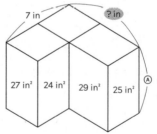

PUZZLE 79

Ⓐ . . . (28 + 27) ÷ 5 = 11 in.
Ⓑ . . . 11 − 3 = 8 in.
Ⓒ . . . 8 + 4 = 12 in.
Area ⑦ is (6 × 12) − 29 = 43 in.²

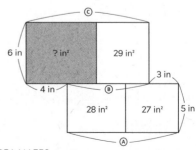

PUZZLE 80

Find the area with stripes:
(8 × 7) − 26 = 30 in.²

This equals the given area of 30, and has the same height.

Therefore it must have the same width.

Length ⑦ is 7 in.

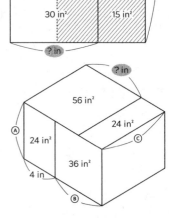

PUZZLE 81

Ⓐ . . . 24 ÷ 4 = 6 in.
Ⓑ . . . 36 ÷ 6 = 6 in.
ⓒ . . . (56 + 24) ÷ (4 + 6) = 8 in.

Length ⑦ is 56 ÷ ⓒ = 7 in.

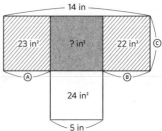

PUZZLE 82

Add the areas with stripes: 23 + 22 = 45 in.²
Consider sides Ⓐ + Ⓑ together: 14 − 5 = 9 in.
ⓒ . . . 45 ÷ 9 = 5 in.

Area ⑦ is ⓒ × 5 = 25 in.²

PUZZLE 83

Add the areas with stripes:
13 + 11 = 24 in.²

This equals the given area of 24, and has the same height.

Therefore Ⓐ must equal Ⓑ.

Similarly:

Area ⑦ is 64 in.²

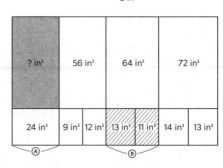

PUZZLE 84

Ⓐ . . . 72 ÷ 9 = 8 in.

Ⓑ . . . 8 – 3 = 5 in.

The two areas of 27 share side Ⓒ, so Ⓓ must also equal Ⓑ: Ⓓ = 5 in.

Length ? is 9 – Ⓓ = 4 in.

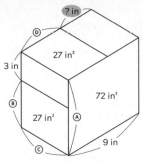

PUZZLE 85

Since 50 is double 25, Ⓐ must be double 4: Ⓐ = 8 in.

Ⓑ . . . 8 + 4 = 12 in.

Add the bottom two areas: 60 + 20 = 80 in.²

This is quadruple 20, so Ⓑ must be quadruple Ⓒ.

Ⓒ . . . 12 ÷ 4 = 3 in.

Length ? is 12 – 3 = 9 in.

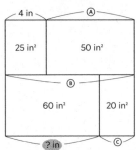

PUZZLE 86

Ⓐ + Ⓒ = 7, and Ⓑ + Ⓒ = 7.

Therefore Ⓐ must equal Ⓑ.

The areas of Ⓓ and 10 are the same height and width: Ⓓ = 10 in.²

Ⓔ + Ⓕ = 8, and Ⓖ + Ⓕ = 8.

Therefore Ⓔ must equal Ⓖ.

Ⓗ and Ⓓ are the same height and width: Ⓗ = 10 in.²

Similarly, area ? is 10 in.²

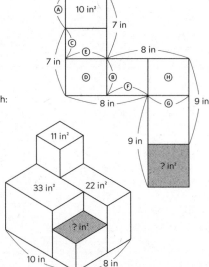

PUZZLE 87

Add the three given areas:
11 + 22 + 33 = 66 in.²

Area ? is (10 × 8) – 66 = 14 in.²

PUZZLE 88

Ⓐ . . . (10 × 6) − 29 = 31 in.²

Ⓐ and Ⓑ are the same height and width:
   Ⓑ = 31 in.²

Ⓒ . . . (11 × 6) − 31 = 35 in.²

Ⓒ and Ⓓ are the same height and width: Ⓓ = 35 in.²

Area ⑦ is (12 × 6) − 35 = 37 in.²

PUZZLE 89

Since 24 is double 12, Ⓐ must be double 15: Ⓐ = 30 in.²

This is triple 10, so 27 must be triple Ⓑ: Ⓑ = 9 in.²

This is half of 18, so 14 must be half of ⑦.

Area ⑦ is 28 in.²

PUZZLE 90

The areas of 40 and 20 share a side, so Ⓐ must be double Ⓑ.

Therefore ⑦ must be double 25.

Area ⑦ is 50 in.²

PUZZLE 91

Ⓐ . . . (6 × 12) − 17 − 15 = 40 in.²

This equals the given area of 40, so Ⓑ must equal 6.

Similarly, Ⓒ must equal 15.

Area ⑦ is (Ⓑ × 13) − 40 − Ⓒ = 23 in.²

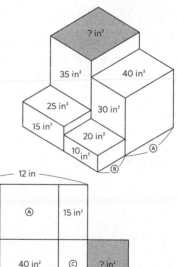

PUZZLE 92

Ⓐ . . . 5 × 4 = 20 in.²

This is double 10, so Ⓑ must be double
11: Ⓑ = 22 in.²

Area ? is Ⓐ + Ⓑ = 42 in.²

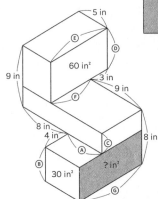

PUZZLE 93

Ⓐ . . . 5 + 9 − 8 = 6 in.

Ⓑ . . . 30 ÷ 6 = 5 in.

Ⓒ . . . 8 − 5 = 3 in.

Ⓓ . . . 9 − 3 = 6 in.

Ⓔ . . . 60 ÷ 6 = 10 in.

Ⓕ . . . 10 − 3 = 7 in.

Ⓖ . . . 4 + 7 = 11 in.

Area ? is Ⓑ × Ⓖ = 55 in.²

PUZZLE 94

On left, the two areas of 20 are the
same width, so Ⓐ must equal Ⓑ.

Therefore Ⓒ must equal 8.

On top, the two areas of 20 are the
same height, so Ⓓ must equal Ⓔ.

Therefore Ⓕ must equal 9.

Area ? is Ⓒ × Ⓕ = 72 in.²

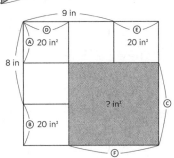

PUZZLE 95

Ⓐ . . . 15 ÷ 5 = 3 in.

This equals the given height of 3, so Ⓑ
must equal Ⓒ.

Therefore, the area with stripes must
equal 21.

Length ? is (21 + 24) ÷ 5 = 9 in.

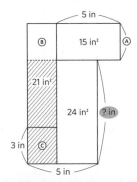

PUZZLE 96

Since 28 is double 14, Ⓐ must be double Ⓑ.

The areas of ⑦ and 16 are the same height, with widths of Ⓐ and Ⓑ. So ⑦ must be double 16.

Area ⑦ is 32 in.²

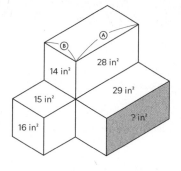

PUZZLE 97

Ⓐ . . . 9 × 5 = 45 in.²

Ⓑ . . . (42 + 49) − 45 = 46 in.²

This equals the given area of 46, so Ⓒ must equal 9.

Length ⑦ is (44 + 46) ÷ Ⓒ = 10 in.

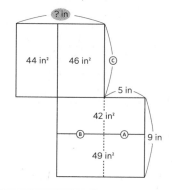

PUZZLE 98

Ⓐ . . . 36 ÷ (5 + 4) = 4 in.

Ⓑ . . . 5 + 4 − 6 = 3 in.

Ⓒ . . . 9 ÷ 3 = 3 in.

Ⓓ . . . 3 + 2 = 5 in.

This equals the given height of 5.

Add the areas with stripes: 14 + 21 = 35 in.²

Find their combined width: 35 ÷ 5 = 7 in.

Length ⑦ is 7 + Ⓐ + Ⓒ = 14 in.

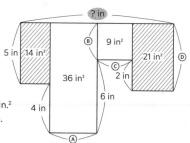

PUZZLE 99

Ⓐ ... 28 ÷ 7 = 4 in.
Ⓑ ... 44 ÷ 4 = 11 in.
Ⓒ ... (21 + 34) ÷ 11 = 5 in.
Ⓓ ... (39 + 21) ÷ 5 = 12 in.
Ⓔ ... 48 ÷ 12 = 4 in.
Ⓕ ... 12 ÷ 4 = 3 in.
Area ⍰ is Ⓓ × Ⓕ = 36 in.²

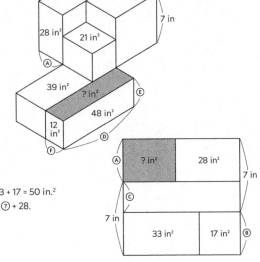

PUZZLE 100

Ⓐ + Ⓒ = 7, and Ⓑ + Ⓒ = 7.
Therefore Ⓐ must equal Ⓑ.
Add the bottom two areas: 33 + 17 = 50 in.²
From Ⓐ = Ⓑ, this must equal ⍰ + 28.
Area ⍰ is 50 − 28 = 22 in.²

PUZZLE 101

Ⓐ ... 3 × 4 = 12 in.²
Ⓑ ... 27 − 12 = 15 in.²
This is half of 30, so 4 must be half of Ⓒ:
  Ⓒ = 8 in.
This is double the height of Ⓓ, so 34
  must be double Ⓓ: Ⓓ = 17 in.²
Area ⍰ is (4 × 3) + Ⓓ = 29 in.²

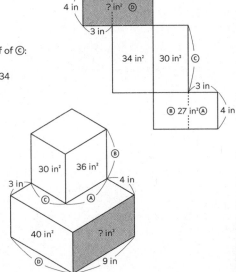

PUZZLE 102

Ⓐ ... 9 − 3 = 6 in.
Ⓑ ... 36 ÷ 6 = 6 in.
Ⓒ ... 30 ÷ 6 = 5 in.
Ⓓ ... 5 + 4 = 9 in.
The areas of 40 and ⍰ are
  the same height and width.
Area ⍰ is 40 in.²

PUZZLE 103

Ⓐ . . . (5 × 13) − 21 − 24 = 20 in.²
Ⓑ . . . 20 ÷ 5 = 4 in.
Ⓒ . . . 6 × 4 = 24 in.²
Length ⑦ is (Ⓒ + 23 + 25) ÷ 6 = 12 in.

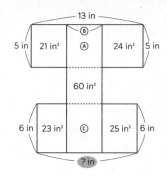

PUZZLE 104

Ⓐ and Ⓑ are the same height and width,
  so they must be equal.
Therefore Ⓑ and the striped area above
  it must total to 24.
Similarly, Ⓒ and Ⓓ must be equal.
Therefore Ⓓ and the striped area above
  it must total to 20.
Find the total area with stripes:
  20 + 24 + 26 = 70 in.²
Ⓔ . . . 70 ÷ 10 = 7 in.
From Ⓒ = Ⓓ, ⑦ must equal Ⓔ.
Length ⑦ is 7 in.

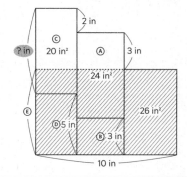

PUZZLE 105

Ⓐ . . . 30 ÷ 6 = 5 in.
Ⓑ . . . 25 ÷ 5 = 5 in.
Ⓒ . . . 35 ÷ 5 = 7 in.
Ⓓ . . . 49 ÷ 7 = 7 in.
Ⓔ = Ⓐ = 5 in.
Ⓕ . . . 40 ÷ 5 = 8 in.
Ⓖ . . . 48 ÷ 8 = 6 in.
Ⓗ . . . 36 ÷ 6 = 6 in.
Ⓘ = Ⓗ = 6 in.
Area ⑦ is Ⓓ × Ⓘ = 42 in.²

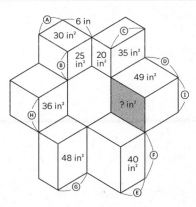

PUZZLE 106

The given areas of 11 are the same
   height, so Ⓐ must equal Ⓑ.
Therefore Ⓒ must equal 22.
Ⓓ . . . 42 − 22 = 20 in.²
Ⓔ . . . 11 − 7 = 4 in.
Ⓑ . . . 20 ÷ 4 = 5 in.
From Ⓐ = Ⓑ, Ⓐ equals 5.
Area ⑦ is (Ⓐ × 7) − 22 = 13 in.²

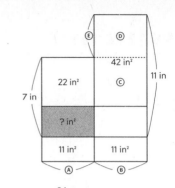

PUZZLE 107

Ⓐ . . . (3 × 8) − 16 = 8 in.²
Ⓑ . . . 21 − 8 = 13 in.²
Since 16 is double Ⓐ, the area with
   stripes must be double Ⓑ.
Area ⑦ is (Ⓑ × 2) − 11 = 15 in.²

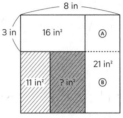

PUZZLE 108

Since 30 is double 15, Ⓐ must be double Ⓑ.
Similarly, Ⓒ must be double 14: Ⓒ = 28 in.²
Ⓐ . . . 28 ÷ 4 = 7 in.
Ⓓ . . . 35 ÷ 7 = 5 in.
Area ⑦ is Ⓓ × 4 = 20 in.²

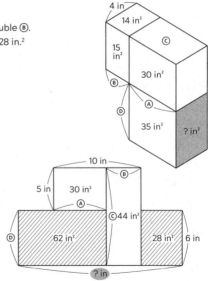

PUZZLE 109

Ⓐ . . . 30 ÷ 5 = 6 in.
Ⓑ . . . 10 − 6 = 4 in.
Ⓒ . . . 44 ÷ 4 = 11 in.
Ⓓ . . . 11 − 5 = 6 in.
This equals the given height of 6.
Add the areas with stripes:
   62 + 28 = 90 in.²
Find their combined width:
   90 ÷ 6 = 15 in.
Length ⑦ is 15 + Ⓑ = 19 in.

PUZZLE 110

(A) . . . (9 × 5) − 34 = 11 in.²

(B) . . . 34 − 11 = 23 in.²

This equals the given area of 23, so (A) must also equal (C): (C) = 11 in.²

(D) . . . 34 − 11 = 23 in.²

(D) and (?) are the same height and width.

Area (?) is 23 in.²

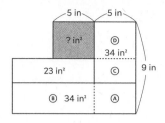

PUZZLE 111

(A) . . . 35 ÷ 5 = 7 in.

(B) . . . 49 ÷ 7 = 7 in.

(C) . . . 42 ÷ 7 = 6 in.

(D) . . . 7 + 6 − 3 = 10 in.

(E) . . . 60 ÷ 10 = 6 in.

From (E) = (C), the areas of 70 and (?) are the same height and width.

Area (?) is 70 in.²

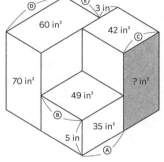

PUZZLE 112

(A) . . . 5 × 2 = 10 in.²

(B) . . . 31 − 10 = 21 in.²

(C) . . . (5 × 7) − 21 = 14 in.²

This equals the given area of 14, so (D) must equal 5.

Therefore (B) and (?) are the same height and width.

Area (?) is 21 in.²

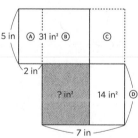

PUZZLE 113

(A) . . . 3 × 7 = 21 in.²

(B) . . . 40 − 21 = 19 in.²

(C) . . . 2 × 7 = 14 in.²

Find the area with stripes: 19 + 14 = 33 in.²

This equals the given area of 33, so it must be the same width.

Length (?) is 7 in.

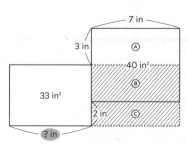

PUZZLE 114

Since 34 is double 17, Ⓐ must be double Ⓑ.
Similarly, Ⓒ must be double 20: Ⓒ = 40 in.²
Ⓓ . . . (20 + 40) ÷ 6 = 10 in.
Ⓔ . . . (17 + 34 + 26 + 13) ÷ 10 = 9 in.
Area ⑦ is 6 × Ⓔ = 54 in.²

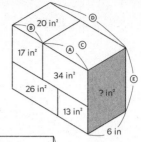

PUZZLE 115

Ⓐ . . . 9 − 4 = 5 in.
Ⓑ . . . 10 − 5 = 5 in.
This equals Ⓐ, so Ⓒ must be half of 44:
  Ⓒ = 22 in.²
Length ⑦ is (22 + 33) ÷ Ⓑ = 11 in.

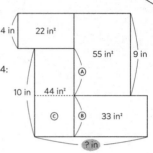

PUZZLE 116

Add the areas with stripes: 32 + 16 = 48 in.²
This is triple 16, so 42 must be triple Ⓐ: Ⓐ = 14 in.²
Ⓑ . . . 42 − 14 = 28 in.²
Length ⑦ is (32 + Ⓑ) ÷ 5 = 12 in.

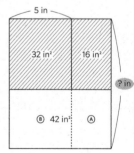

PUZZLE 117

Ⓐ . . . 24 ÷ 4 = 6 in.
Ⓑ . . . 42 ÷ 6 = 7 in.
Ⓒ . . . 35 ÷ 7 = 5 in.
Ⓓ . . . 30 ÷ 5 = 6 in.
Ⓔ . . . 36 ÷ 6 = 6 in.
Ⓕ . . . 48 ÷ 6 = 8 in.
The areas of ⑦ and 54 share side Ⓖ,
  and 4 is half of Ⓕ.
Therefore ⑦ must be half of 54.
Area ⑦ is 27 in.²

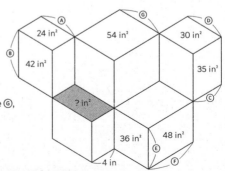

PUZZLE 118

The given areas of 10 are all the same width, so they must be the same height.

Therefore Ⓐ and Ⓑ must both equal 25.

Ⓒ . . . 53 − 25 = 28 in.²

The area with stripes must also equal 28.

Ⓓ . . . 28 ÷ (3 + 4) = 4 in.

Ⓔ . . . 4 × 3 = 12 in.²

Area ⑦ is Ⓐ + Ⓔ = 37 in.²

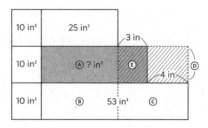

PUZZLE 119

Since 39 is triple 13, 42 must be triple Ⓐ:
Ⓐ = 14 in.²

This equals 56 ÷ 4, so Ⓑ must equal
64 ÷ 4: Ⓑ = 16 in.²

This equals 80 ÷ 5, so Ⓒ must equal
35 ÷ 5: Ⓒ = 7 in.²

This is half of Ⓐ, so Ⓓ must be half of
42: Ⓓ = 21 in.²

Since 63 is triple 21, ⑦ must be triple 25.

Area ⑦ is 75 in.²

	35 in²		80 in²	
42 in²	Ⓓ	25 in²		39 in²
	63 in²	? in²		
56 in²			64 in²	
Ⓐ	Ⓒ		Ⓑ	13 in²

PUZZLE 120

Draw the area with stripes: 7 × 6 = 42 in.²

The areas of 36 and 18 share a side, and
36 is double 18.

Similarly, 42 must be double ⑦.

Area ⑦ is 21 in.²

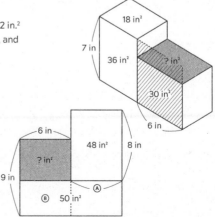

PUZZLE 121

Ⓐ . . . 48 ÷ 8 = 6 in.

This equals the given width
of 6, so Ⓑ must be half of
50: Ⓑ = 25 in.²

Area ⑦ is (9 × 6) − Ⓑ = 29 in.²

PUZZLE 122

(A)... 9 − 4 = 5 in.

(B)... 3 + 5 = 8 in.

This is double 4, so 30 must be double (C): (C) = 15 in.²

(D)... 39 − 15 = 24 in.²

(E)... 24 ÷ 4 = 6 in.

Area (?) is (A) × (E) = 30 in.²

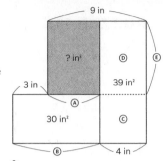

PUZZLE 123

The given areas of 20 are the same height, so (A) must equal (B).

(A) × (B) = 25, so (A) and (B) = 5 in.

Length (?) is 20 ÷ (B) = 4 in.

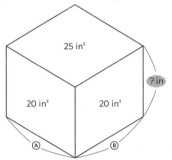

PUZZLE 124

(A)... 48 ÷ 8 = 6 in.

(B)... 9 − 6 = 3 in.

This is half of (A), so 14 must be half of (C): (C) = 28 in.²

(D)... 48 − 28 = 20 in.²

Length (?) is (22 + 20) ÷ (A) = 7 in.

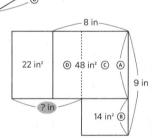

PUZZLE 125

(A)... 7 × 6 = 42 in.²

This is double 21, so 58 must be double (B): (B) = 29 in.²

(C)... 50 − 29 = 21 in.²

This equals the given area of 21 and has the same height, so it must have the same width.

Length (?) is 6 in.

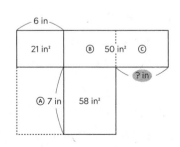

PUZZLE 126

(A) . . . 20 ÷ 5 = 4 in.

(B) . . . 9 – 4 = 5 in.

The areas of (C) and 29 share a side,
    and their other sides both equal 5:
    (C) = 29 in.^{2}

(D) . . . (29 + 21) ÷ 5 = 10 in.

Area ⑦ is (A) × (D) = 40 in.^{2}

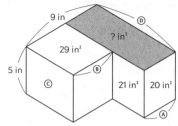

PUZZLE 127

Since 56 is double 28, (A) must be double (B).
Therefore (C) must be double 17: (C) = 34 in.^{2}

Length ⑦ is ((C) + 56) ÷ 9 = 10 in.

PUZZLE 128

Since 52 is double 26, (A) must be double
    27: (A) = 54 in.^{2}

(B) . . . 80 – 54 = 26 in.^{2}

This is half of 52, so 20 must be half of ⑦.

Area ⑦ is 40 in.^{2}

PUZZLE 129

The given areas of 34 share a side, so
    (A) must equal (B).

(C) shares side ⑦ with a given area of
    35, and from (A) = (B), their other
    sides are also equal. So (C) must
    equal 35.

(C) is the same height as the other
    given area of 35, so ⑦ must equal (D).

The area of 36 facing up equals ⑦ × (D).
    From ⑦ = (D), both must be 6.

Length ⑦ is 6 in.

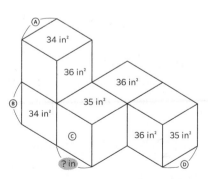

PUZZLE 130

The areas of Ⓐ and 38 are the same height and width: Ⓐ = 38 in.²

Ⓑ . . . 77 − 38 = 39 in.²

This equals the given area of 39, so Ⓐ must equal Ⓒ: Ⓒ = 38 in.²

Ⓓ . . . 68 − 38 = 30 in.²

Length ⑦ is Ⓓ ÷ 6 = 5 in.

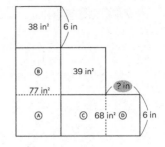

PUZZLE 131

Ⓐ . . . (4 × 7) − 15 = 13 in.²

Ⓑ . . . 13 − 7 = 6 in.

Ⓒ . . . 6 × 4 = 24 in.²

Add the areas with stripes:
24 + 15 = 39 in.²

This is triple Ⓐ, so ⑦ must be triple 14.

Area ⑦ is 42 in.²

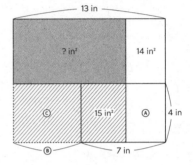

PUZZLE 132

Ⓐ . . . (6 × 10) − 40 = 20 in.²

This is half of 40, so Ⓑ must be double Ⓒ.

Therefore Ⓓ must be double 17: Ⓓ = 34 in.²

Also, 26 must be double Ⓔ: Ⓔ = 13 in.²

Add the areas facing left: 17 + 34 + 13 + 26 = 90 in.²

Length ⑦ is 90 ÷ 10 = 9 in.

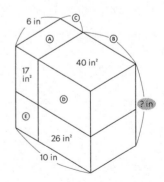

PUZZLE 133

Ⓐ . . . 20 ÷ 4 = 5 in.

Add the areas with stripes: 19 + 21 = 40 in.²

Find their combined height: 13 − 5 = 8 in.

Ⓑ . . . 40 ÷ 8 = 5 in.

Similarly, consider the areas of 22 and 23 together.

Ⓒ . . . (22 + 23) ÷ (14 − 5) = 5 in.

Area Ⓐ is Ⓑ × Ⓒ = 25 in.²

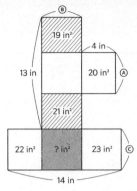

PUZZLE 134

Since 24 is double 12, Ⓐ must be double 21: Ⓐ = 42 in.²

Since 11 is half of 22, ⑦ must be half of Ⓐ.

Area ⑦ is 21 in.²

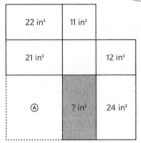

PUZZLE 135

Ⓐ . . . 4 + 4 = 8 in.

Ⓑ . . . 56 ÷ 8 = 7 in.

Add the areas with stripes: 32 + 24 = 56 in.²

Ⓒ . . . 56 ÷ (15 − 7) = 7 in.

Area ⑦ is Ⓑ × Ⓒ = 49 in.²

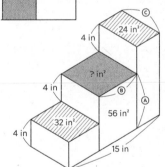

PUZZLE 136

Ⓐ . . . (8 × 9) − 19 − 34 = 19 in.²

This equals the given area of 19, so Ⓑ must equal Ⓒ.

Therefore Ⓓ must be half of 34: Ⓓ = 17 in.²

Ⓑ + Ⓒ = 8, so from Ⓑ = Ⓒ, both equal 4.

Length ⑦ is (39 + 17) ÷ Ⓒ = 14 in.

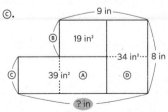

PUZZLE 137

Add the given areas: 31 + 21 + 28 = 80 in.²
(A) . . . (9 × 10) − 80 = 10 in.²
Add the areas with stripes: 10 + 21 = 31 in.²
This equals the given area of 31, so (B) must
be half of 28: (B) = 14 in.²
Also, (C) must be half of 10: (C) = 5 in.
Length (?) is (21 + (B)) ÷ (C) = 7 in.

PUZZLE 138

Since 46 is double 23, (A) must be double (B).
The areas of (C) and 24 share a side, and their
other sides are (A) and (B). So (C) must be
double 24: (C) = 48 in.²
This is triple 16, so (D) must be triple (E).
The areas of 42 and (?) are the same height,
with widths of (D) and (E). So 42 must be triple (?).
Area (?) is 14 in.²

PUZZLE 139

Since 54 is triple 18, (A) must be triple 19:
(A) = 57 in.²
(B) . . . 84 − 57 = 27 in.²
This is half of 54, so 4 must be half of (?).
Length (?) is 8 in.

PUZZLE 140

(A) . . . (7 × 6) − 16 = 26 in.²
(B) . . . 39 − 26 = 13 in.²
(A) is double (B), so the area with stripes
must be double 28: 56 in.²
Length (?) is 56 ÷ 7 = 8 in.

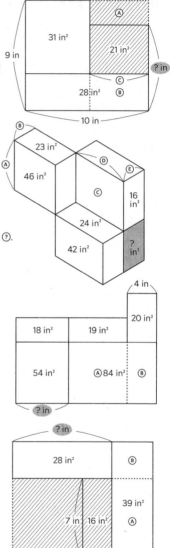

PUZZLE 141

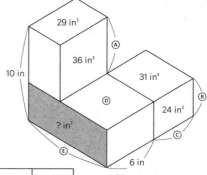

Ⓐ . . . 36 ÷ 6 = 6 in.

Ⓑ . . . 10 – 6 = 4 in.

Ⓒ . . . 24 ÷ 4 = 6 in.

The areas of 31 and Ⓓ share a side, and their other sides both equal 6. So Ⓓ must equal 31.

Ⓔ . . . (29 + 31) ÷ 6 = 10 in.

Area ⑦ is Ⓑ × Ⓔ = 40 in.²

PUZZLE 142

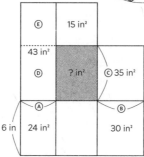

Ⓐ . . . 24 ÷ 6 = 4 in.

Ⓑ . . . 30 ÷ 6 = 5 in.

Ⓒ . . . 35 ÷ 5 = 7 in.

Ⓓ . . . 7 × 4 = 28 in.²

Ⓔ . . . 43 – 28 = 15 in.²

This equals the given area of 15, so Ⓓ must equal ⑦.

Area ⑦ is 28 in.²

PUZZLE 143

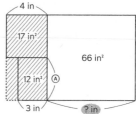

Ⓐ . . . 12 ÷ 3 = 4 in.

Find the area with stripes: 17 + (4 × 4) = 33 in.²

This is half of 66, so 4 must be half of ⑦.

Length ⑦ is 8 in.

PUZZLE 144

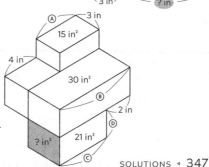

Ⓐ . . . 15 ÷ 3 = 5 in.

Ⓑ . . . 4 + 5 = 9 in.

Ⓒ . . . 9 – 2 = 7 in.

Ⓓ . . . 21 ÷ 7 = 3 in.

The areas of 30 and ⑦ share a side, and their other sides are Ⓑ and Ⓓ.

Ⓑ is triple Ⓓ, so 30 must be triple ⑦.

Area ⑦ is 10 in.²

PUZZLE 145

Ⓐ . . . 21 ÷ 3 = 7 in.

This equals the given height of 7.

Add the three areas with height 7:
29 + 21 + 27 = 77 in.^{2}

Find their combined width: 77 ÷ 7 = 11 in.

Length ⑦ is 11 in.

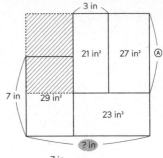

PUZZLE 146

Ⓐ . . . (9 × 7) – 25 = 38 in.^{2}

Ⓑ . . . 57 – 38 = 19 in.^{2}

Since 57 is triple Ⓑ, 48 must
be triple Ⓒ: Ⓒ = 16 in.^{2}

Ⓓ . . . 48 – 16 = 32 in.^{2}

Length ⑦ is (Ⓐ + Ⓓ) ÷ 7 = 10 in.

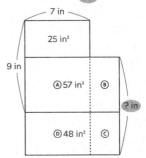

PUZZLE 147

Since 52 is double 26, Ⓐ must be double Ⓑ.

Since 34 is double 17, Ⓒ must be double Ⓓ.

Ⓑ and Ⓓ are the same height, so from
the above, Ⓐ must equal Ⓒ.

Therefore the areas of 65 and ⑦
are the same height and width.

Area ⑦ is 65 in.^{2}

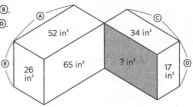

PUZZLE 148

Add the areas with stripes:
19 + 25 = 44 in.^{2}

Find their combined width: 15 – 5 = 10 in.

This is double 5, so the area with stripes
must be double Ⓐ: Ⓐ = 22 in.^{2}

Area ⑦ is (13 × 5) – 24 – Ⓐ = 19 in.^{2}

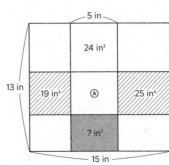

PUZZLE 149

Ⓐ . . . 10 − 5 = 5 in.

This is half of 10, so Ⓑ must equal 21.

Ⓒ . . . (21 + 21 + 58) ÷ 10 = 10 in.

Ⓓ . . . 10 − 5 = 5 in.

This is half of Ⓒ, so Ⓔ must equal 24.

Ⓕ . . . (52 + 24 + 24) ÷ 10 = 10 in.

Ⓖ . . . 10 − 5 = 5 in.

This is half of Ⓕ, so Ⓗ must equal Ⓟ.

Area Ⓟ is (Ⓕ × 10 − 54) ÷ 2 = 23 in.²

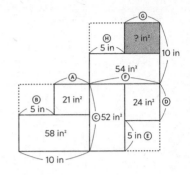

PUZZLE 150

Ⓐ . . . (9 × 5) − 24 = 21 in.²

Ⓑ . . . 45 − 21 = 24 in.²

This is half of 48, so Ⓒ must be half of Ⓓ.

Ⓐ and Ⓔ share a side, and their other sides are Ⓒ and Ⓓ. So Ⓐ must be half of Ⓔ:
Ⓔ = 42 in.²

Ⓕ . . . 80 − 42 = 38 in.²

Ⓕ and Ⓟ share a side, and their other sides are Ⓒ and Ⓓ. So Ⓟ must be half of Ⓕ.

Area Ⓟ is 19 in.²

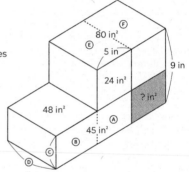

PUZZLE 151

Add the bottom areas: 27 + 45 = 72 in.²

Ⓐ . . . 4 + 7 + 5 = 16 in.

This is quadruple 4, so 72 must be quadruple Ⓑ: Ⓑ = 18 in.²

Ⓒ . . . 27 − 18 = 9 in.²

This is half of Ⓑ, so Ⓓ must be half of 4:
Ⓓ = 2 in.

Length Ⓟ is 7 − Ⓓ + 5 = 10 in.

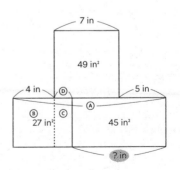

PUZZLE 152

Ⓐ . . . 20 ÷ 5 = 4 in.

Ⓑ . . . 4 + 3 = 7 in.

Find the area with stripes: 7 × 6 = 42 in.²

This is double 21, so 18 must be double Ⓐ.

Area Ⓐ is 9 in.²

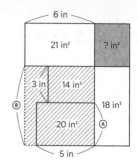

PUZZLE 153

Ⓐ . . . 16 ÷ 4 = 4 in.

The areas of 15 and Ⓑ share a side,
and their other sides both equal 4.
So Ⓑ must equal 15.

Ⓒ . . . 35 − 15 = 20 in.²

Ⓓ . . . 20 ÷ 4 = 5 in.

Ⓔ . . . 25 ÷ 5 = 5 in.

Draw area Ⓕ that shares a side with
the area of 24. Their other sides,
Ⓓ and Ⓔ, are equal: Ⓕ = 24 in.²

Ⓖ . . . 54 − 24 = 30 in.²

Ⓗ . . . 30 ÷ 5 = 6 in.

Ⓘ . . . 36 ÷ 6 = 6 in.

This equals Ⓗ, so the areas of 45 and Ⓐ are the same height and width.

Area Ⓐ is 45 in.²

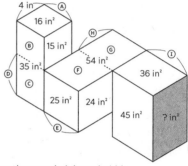

PUZZLE 154

Ⓐ . . . 6 × 5 = 30 in.²

Ⓑ . . . 53 − 30 = 23 in.²

This equals the given area of 23, so Ⓒ
must equal 25.

Ⓓ . . . 48 − 25 = 23 in.²

This equals the given area of 23, so it
must be the same height.

Length Ⓐ is 6 in.

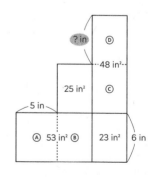

PUZZLE 155

Consider the areas on top:
  12 + 15 = 27 in.²
  15 + 39 = 54 in.²

This is double 27, so Ⓐ must be double 7:
  Ⓐ = 14 in.²

Similarly, 16 + 40 is double 12 + 16, so Ⓑ
  must be double 9: Ⓑ = 18 in.

Area ⑦ is Ⓐ × Ⓑ = 252 in.²

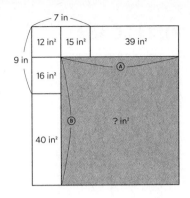

PUZZLE 156

Since 30 is double 15, Ⓐ must be double Ⓑ.

Area Ⓒ shares side Ⓓ with the area of 25 on
  top, and their other sides are Ⓐ and Ⓑ.

So, Ⓒ must be double 25: Ⓒ = 50 in.²

This is double the adjacent area of 25,
  so Ⓓ must be double Ⓔ.

Therefore ⑦ must be double 20.

Area ⑦ is 40 in.²

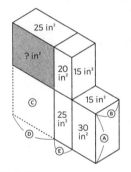

PUZZLE 157

Ⓐ . . . 6 × 7 = 42 in.²

This is double 21, so Ⓑ must be double Ⓒ.

Therefore ⑦ + 56 must be double 34 + 21.

(34 + 21) × 2 = 110 in.²

Area ⑦ is 110 − 56 = 54 in.²

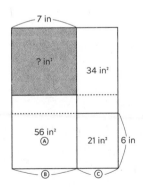

PUZZLE 158

Ⓐ . . . 36 ÷ 9 = 4 in.

Ⓑ . . . 24 ÷ (10 − 4) = 4 in.

Ⓒ . . . 9 − 4 = 5 in.

Ⓓ . . . 13 − 5 = 8 in.

Ⓔ . . . 32 ÷ 8 = 4 in.

Ⓕ . . . 21 ÷ (11 − 4) = 3 in.

Ⓖ . . . 8 − 3 = 5 in.

This equals Ⓒ, so the area with stripes must equal Ⓓ.

Find the total area with height Ⓒ: (4 + 10) × 5 = 70 in.²

Find the area with stripes: 70 − 43 = 27 in.²

Area ⑦ is 27 in.²

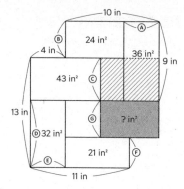

PUZZLE 159

Since 54 is double 27, Ⓐ must be double 5: Ⓐ = 10 in.

Ⓑ . . . 30 ÷ 5 = 6 in.

Ⓒ . . . 10 − 6 = 4 in.

Since 50 is double 25, Ⓓ must be double Ⓒ: Ⓓ = 8 in.

Ⓔ . . . 12 ÷ 4 = 3 in.

Ⓕ . . . 8 − 3 = 5 in.

Since 48 is double 24, Ⓖ must be double Ⓕ: Ⓖ = 10 in.

Area ⑦ is Ⓕ × Ⓖ = 50 in.²

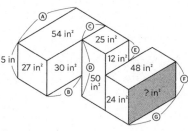

PUZZLE 160

Ⓐ . . . 4 × 5 = 20 in.²

Ⓑ . . . 34 − 20 = 14 in.²

Find the area with stripes: 28 + 14 = 42 in.²

This is double 21, so 4 must be double Ⓒ: Ⓒ = 2 in.

Area ⑦ is (3 + Ⓒ) × 5 = 25 in.²

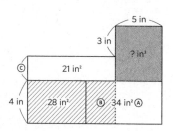

PUZZLE 161

Since 54 is triple 18, 7 must be triple Ⓐ.
Since 34 is double 17, Ⓐ must be double Ⓑ.
Therefore 7 is sextuple Ⓑ.
Since 48 is triple 16, Ⓒ must be triple Ⓑ.
Since 52 is double 26, ? must be double Ⓒ.
Therefore ? is also sextuple Ⓑ.
Length ? is 7 in.

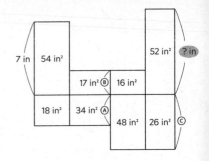

PUZZLE 162

Ⓐ . . . (24 + 26) ÷ 10 = 5 in.
Ⓑ . . . 10 − 5 = 5 in.
Ⓒ . . . 25 ÷ 5 = 5 in.
Ⓓ . . . 10 − 5 = 5 in.

The areas of ? and 26 share a side, and
their other sides, Ⓐ and Ⓓ, are equal.

Area ? is 26 in.²

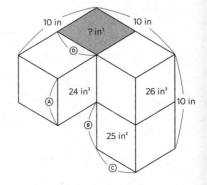

PUZZLE 163

Ⓐ . . . (6 × 3) − 10 = 8 in.²
Ⓑ . . . 24 − 8 = 16 in.²
This is double Ⓐ, so Ⓒ must
be double 3: Ⓒ = 6 in.
Also, Ⓓ must be double 10: Ⓓ = 20 in.²
Ⓔ . . . 42 − 20 = 22 in.²
Ⓕ . . . 6 + 3 = 9 in.
This is triple 3, so 30 must be triple Ⓖ: Ⓖ = 10 in.²
Ⓗ . . . 30 − 10 = 20 in.²
Length ? is (Ⓔ + Ⓗ) ÷ Ⓒ = 7 in.

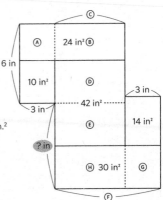

PUZZLE 164

Find the area with stripes: (5 × 5) + 21 = 46 in.²
This equals the given area of 46, so Ⓐ must equal 5.
Ⓑ . . . 46 − (5 × 4) = 26 in.²
Ⓒ . . . 5 × 6 = 30 in.²
Add Ⓑ + Ⓒ: 26 + 30 = 56 in.²
This equals the given area of 56, so Ⓓ must equal Ⓐ.
Ⓔ . . . 56 − (5 × 4) = 36 in.²
This equals the given area of 36, so Ⓐ must equal Ⓓ.
Length Ⓐ is 5 in.

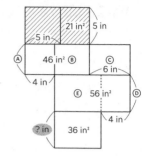

PUZZLE 165

Ⓐ . . . 16 ÷ 4 = 4 in.
Find the area with stripes: 9 + 25 = 34 in.²
Find Ⓑ + Ⓒ together: 12 − 4 = 8 in.
This is double Ⓐ.
The area with stripes and Ⓐ share a side,
 and their other sides are Ⓑ + Ⓒ and Ⓐ.
Therefore the area with stripes must be double Ⓐ.
Area Ⓐ is 34 ÷ 2 = 17 in.²

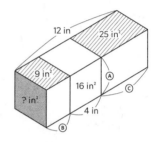

PUZZLE 166

Draw area Ⓐ equal to the given area of 24.
They must have equal width, so Ⓑ equals 8.
Ⓒ . . . 49 − 24 = 25 in.²
Ⓓ . . . 12 − 8 = 4 in.
Ⓑ is double Ⓓ, so the area with stripes
 must be double Ⓒ: 50 in.²
Area Ⓐ is 50 − Ⓐ − 16 = 10 in.²

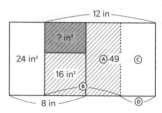

PUZZLE 167

Since 30 is triple 10, Ⓐ must be triple Ⓑ.
Therefore 16 + Ⓒ must be triple 12.
Ⓒ . . . (12 × 3) − 16 = 20 in.²
Add the areas with stripes: 20 + 12 = 32 in.²
This is double 16, so Ⓐ must be double 10.
Area Ⓐ is 20 in.²

PUZZLE 168

Since 28 is double 14, Ⓐ must be double Ⓑ.

Therefore the area with stripes must be double 15.

This equals the given area of 30, and they share side Ⓒ. So their other sides must be equal: Ⓐ = Ⓓ.

Ⓐ × Ⓓ = 49, so Ⓐ and Ⓓ equal 7.

Length ⑦ is 28 ÷ Ⓐ = 4 in.

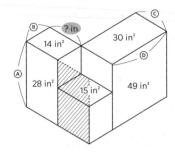

PUZZLE 169

The given widths of 10 are equal, so Ⓐ must equal Ⓑ.

Therefore Ⓒ must equal the given area of 17.

Add the upper left areas: 17 + 43 = 60 in.²

Add the upper right areas: 13 + 17 = 30 in.²

This is half of 60, so ⑦ must be half of 58.

Area ⑦ is 29 in.²

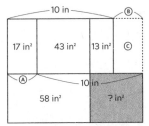

PUZZLE 170

Ⓐ . . . 30 ÷ 6 = 5 in.

This equals the given height of 5, so the upper area with stripes must equal 17.

Ⓑ . . . (17 + 18) ÷ 5 = 7 in.

Ⓒ . . . 49 ÷ 7 = 7 in.

Find the lower area with stripes: (7 × 6) − 25 = 17 in.²

This is the same height as a given area of 17, so it must be the same width.

Length ⑦ is 6 in.

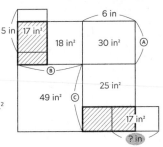

PUZZLE 171

Since 48 is double 24, Ⓐ must be double 5: Ⓐ = 10 in.

Ⓑ . . . 5 × 10 = 50 in.²

This is double 25, so Ⓒ must be double Ⓓ.

Therefore 26 must be double ⑦.

Area ⑦ is 13 in.²

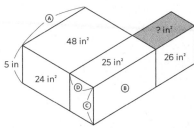

PUZZLE 172

Ⓐ . . . 42 ÷ 6 = 7 in.

This equals the given width of 7,
 so Ⓑ must equal 27.

Ⓒ . . . 36 − 27 = 9 in.²

Since 27 is triple 9, 42 must be triple ⑦.

Area ⑦ is 14 in.²

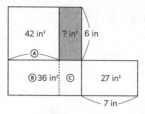

PUZZLE 173

Add the given areas on top: 18 + 40 = 58 in.²

Find the area with stripes: (10 × 9) − 58 = 32 in.²

This is double 16, so 58 must be
 double (⑦ + 18).

Area ⑦ is (58 ÷ 2) − 18 = 11 in.²

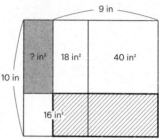

PUZZLE 174

Ⓐ . . . 40 ÷ 8 = 5 in.

Ⓑ . . . (5 × 7) − 24 = 11 in.²

Ⓒ . . . 40 − 11 = 29 in.²

The areas of Ⓒ and 29 are equal, and
 they share a side. So Ⓓ must equal Ⓐ.

Area ⑦ is Ⓓ × 7 = 35 in.²

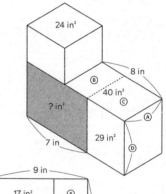

PUZZLE 175

Ⓐ . . . (3 × 9) − 17 = 10 in.²

Ⓑ . . . 30 − 10 = 20 in.²

This is double Ⓐ, so Ⓒ must be double 3:
 Ⓒ = 6 in.

Ⓓ . . . 13 − 3 − 6 = 4 in.

Ⓔ . . . 28 ÷ 4 = 7 in.

Area ⑦ is (Ⓒ × Ⓔ) − Ⓑ = 22 in.²

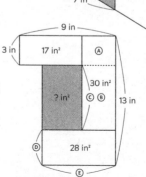

PUZZLE 176

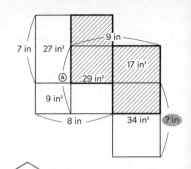

Find the upper area with stripes:
  $(7 \times 8) - 27 = 29$ in.²

This equals the given area of 29,
  so Ⓐ must equal 7.

Find the lower area with stripes:
  $(7 \times 9) - 29 = 34$ in.²

This equals the given area of 34,
  so ⑦ must equal Ⓐ.

Length ⑦ is 7 in.

PUZZLE 177

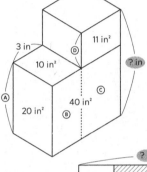

Since 20 is double 10, Ⓐ must
  be double 3: Ⓐ = 6 in.

Ⓑ . . . $6 \times 3 = 18$ in.²

Ⓒ . . . $40 - 18 = 22$ in.²

This is double 11, so Ⓐ must
  be double Ⓓ: Ⓓ = 3 in.

Length ⑦ is Ⓐ + Ⓓ = 9 in.

PUZZLE 178

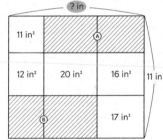

Add the given areas below Ⓐ: $20 + 16 = 36$ in.²

This is triple 12, so Ⓐ must be triple 11: Ⓐ = 33 in.²

Add the given areas above Ⓑ: $12 + 20 = 32$ in.²

This is double 16, so Ⓑ must be double 17: Ⓑ = 34 in.²

Find the total area:
  $11 + 33 + 12 + 20 + 16 + 34 + 17 = 143$ in.²

Length ⑦ is 143 ÷ 11 = 13 in.

PUZZLE 179

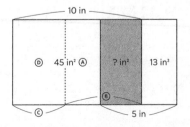

Draw area Ⓐ equal to the given area of 13.

They must have equal width,
  so Ⓑ must equal 5.

Ⓒ . . . $10 - 5 = 5$ in.

This equals Ⓑ, so Ⓐ + ⑦ must equal Ⓓ.

Ⓓ . . . $45 - 13 = 32$ in.²

Area ⑦ is Ⓓ − Ⓐ = 19 in.²

PUZZLE 180

Ⓐ . . . 4 × 8 = 32 in.²
Ⓑ . . . 50 − 32 = 18 in.²
Ⓒ . . . 3 × 8 = 24 in.²
Add Ⓑ and Ⓒ: 18 + 24 = 42 in.²
This equals the given area of 42,
   so Ⓓ must equal 8.
This is double 4, so 46 must be double ⑦.
Area ⑦ is 23 in.²

PUZZLE 181

Ⓐ . . . (3 × 7) − 10 = 11 in.²
Ⓑ . . . 27 − 11 = 16 in.²
This is half of 32, so Ⓐ must be half of
   the area with stripes.
Find the area with stripes: 11 × 2 = 22 in.²
Length ⑦ is (22 + Ⓐ) ÷ 3 = 11 in.

PUZZLE 182

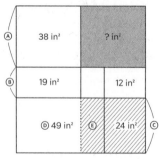

Since 38 is double 19, Ⓐ must be double Ⓑ.
Since 24 is double 12, Ⓒ must also be double Ⓑ.
Ⓐ = Ⓒ, so Ⓓ must equal 38.
Ⓔ . . . 49 − 38 = 11 in.²
Add the areas with stripes: 11 + 24 = 35 in.²
From Ⓐ = Ⓒ, this must equal ⑦.
Area ⑦ is 35 in.²

PUZZLE 183

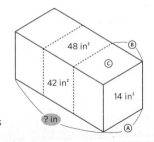

Since 42 is triple 14, ⑦ must be triple Ⓐ.
Divide the area of 48 into three equal areas.
⑦ is also triple Ⓑ, so Ⓐ must equal Ⓑ.
Ⓒ . . . 48 ÷ 3 = 16 in.²
Ⓐ × Ⓑ = 16, so from Ⓐ = Ⓑ, both must equal 4.
Length ⑦ is Ⓑ × 3 = 12 in.

PUZZLE 184

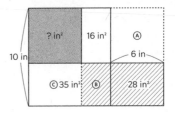

Ⓐ . . . (10 × 6) − 28 = 32 in.²

This is double 16, so 28 must be double Ⓑ:
   Ⓑ = 14 in.²

Ⓒ . . . 35 − 14 = 21 in.²

Add the areas with stripes: 14 + 28 = 42 in.²

This is double Ⓒ, so Ⓐ + 16 must be double ⑦.

Ⓐ + 16 = 48 in.²

Area ⑦ is 24 in.²

PUZZLE 185

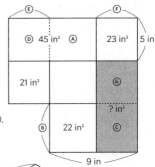

Ⓐ . . . (5 × 9) − 23 = 22 in.²

This equals the given area of 22,
   so Ⓑ must equal 5.

Therefore Ⓒ must equal 23.

Ⓓ . . . 45 − 22 = 23 in.²

This equals the given area of 23, so Ⓔ must equal Ⓕ.

Therefore Ⓖ must equal 21.

Area ⑦ is Ⓒ + Ⓖ = 44 in.²

PUZZLE 186

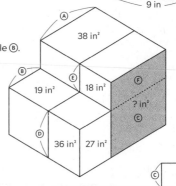

Since 38 is double 19, Ⓐ must be double Ⓑ.

Therefore Ⓒ must be double 27:
   Ⓒ = 54 in.²

Since 36 is double 18, Ⓓ must be
   double Ⓔ.

Therefore Ⓒ must be double Ⓕ:
   Ⓕ = 27 in.²

Area ⑦ is Ⓒ + Ⓕ = 81 in.²

PUZZLE 187

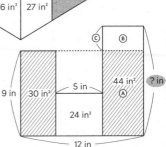

Find the area with stripes: 9 × (12 − 5) = 63 in.²

Ⓐ . . . 63 − 30 = 33 in.²

Ⓑ . . . 44 − 33 = 11 in.²

Since 33 is triple 11, 9 must be triple Ⓒ: Ⓒ = 3 in.

Length ⑦ is 9 + Ⓒ = 12 in.

PUZZLE 188

Ⓐ . . . (4 × 10) − 17 = 23 in.²

Ⓑ . . . (4 × 9) − 23 = 13 in.²

This equals the given area of 13,
   so Ⓒ must equal 4.

Therefore ⑦ must equal Ⓐ.

Area ⑦ is 23 in.²

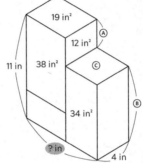

PUZZLE 189

Ⓐ . . . 12 ÷ 4 = 3 in.

Ⓑ . . . 11 − 3 = 8 in.

The areas of Ⓒ and 34
   share a side, and their
   other sides are 4 and Ⓑ.

Ⓑ is double 4, so 34 must
   be double Ⓒ: Ⓒ = 17 in.²

Length ⑦ is (19 + Ⓒ) ÷ 4 = 9 in.

PUZZLE 190

Ⓐ . . . (10 × 5) − 23 = 27 in.²

Ⓑ . . . 14 − 5 − 4 = 5 in.

This equals the given width of 5, so Ⓒ
   must equal Ⓐ: Ⓒ = 27 in.²

Ⓓ . . . 67 − 27 = 40 in.²

Ⓔ . . . 40 ÷ 5 = 8 in.

Area ⑦ is Ⓔ × 4 = 32 in.²

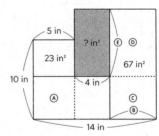

PUZZLE 191

Find the area with stripes: (7 × 10) − 31 = 39 in.²

This equals the given area of 39,
   so Ⓐ must equal 7.

Therefore Ⓑ must equal Ⓒ.

Add these two areas: 47 + 31 = 78 in.²

This is double the area with stripes, so Ⓓ
   must be double Ⓔ.

Therefore 28 must be double Ⓕ: Ⓕ = 14 in.²

From Ⓑ = Ⓒ, ⑦ must equal Ⓕ.

Area ⑦ is 14 in.²

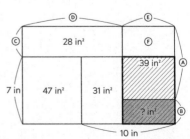

PUZZLE 192

Ⓐ . . . (5 × 9) − 24 = 21 in.²

This equals the given area of 21, so Ⓑ must equal 5.

Ⓒ . . . 5 × 7 = 35 in.²

Ⓓ . . . 56 − 35 = 21 in.²

This equals the given area of 21, so Ⓔ must equal Ⓕ.

Ⓖ . . . 48 − 21 = 27 in.²

Ⓖ and ⓘ share a side, and their other sides, Ⓔ and Ⓕ, are equal.

Area ? is 27 in.²

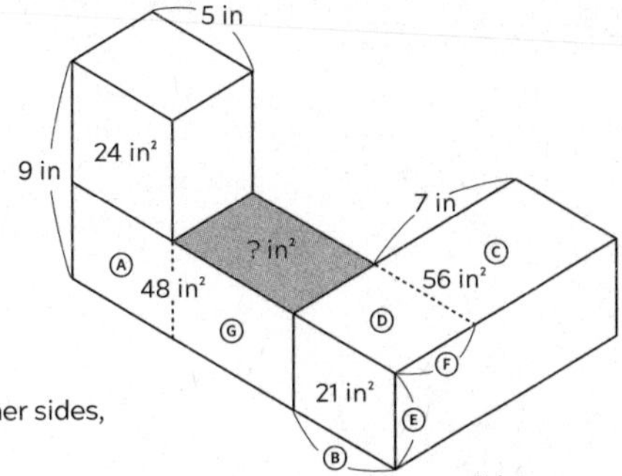

PUZZLE 193

Find the area with stripes: (9 × 11) − 39 = 60 in.²

This is triple 20, so 39 must be triple Ⓐ: Ⓐ = 13 in.²

Find the total area with width ?: 30 + 13 + 20 = 63 in.²

Length ? is 63 ÷ 9 = 7 in.

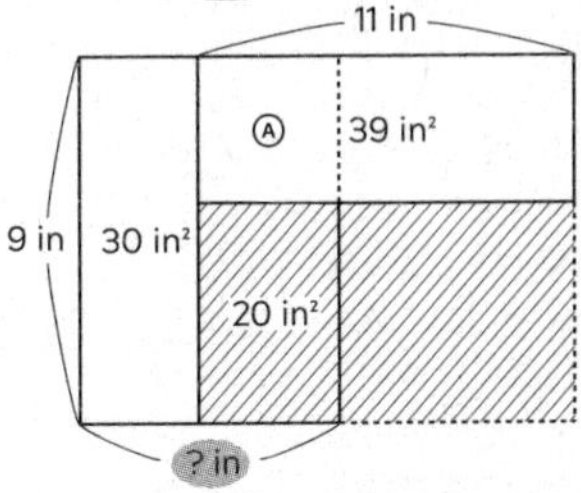

PUZZLE 194

Ⓐ . . . 5 × 4 = 20 in.²

This is double 10, so 24 must be double Ⓑ: Ⓑ = 12 in.²

Ⓒ . . . 30 − 12 = 18 in.²

Add the areas with stripes: 12 + 24 = 36 in.²

This is double Ⓒ, so Ⓐ + 10 must be double Ⓓ: Ⓓ = 15 in.²

Ⓔ . . . 27 − 15 = 12 in.²

Length ? is Ⓔ ÷ 4 = 3 in.

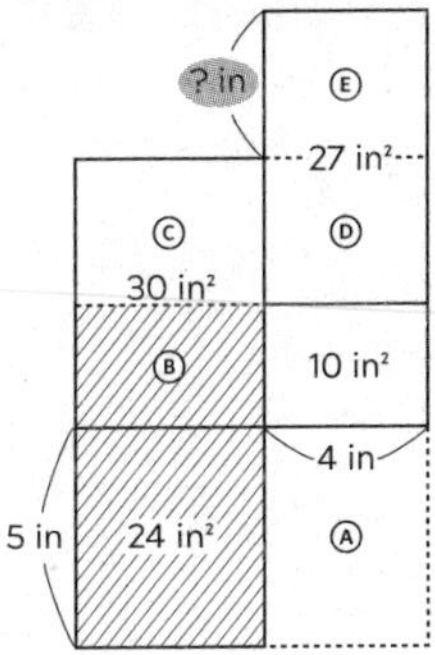

PUZZLE 195

Ⓐ . . . 35 ÷ 7 = 5 in.

Ⓑ . . . 30 ÷ 5 = 6 in.

Ⓒ . . . 28 ÷ 7 = 4 in.

Ⓓ . . . 6 × 4 = 24 in.²

Ⓔ . . . 56 − 24 = 32 in.²

This equals the given area of 32, so Ⓕ must equal Ⓑ.

Ⓖ . . . 18 ÷ 6 = 3 in.

The areas of 32 and ? share a side, and Ⓕ is double Ⓖ.

Therefore 32 must be double ?.

Area ? is 16 in.²

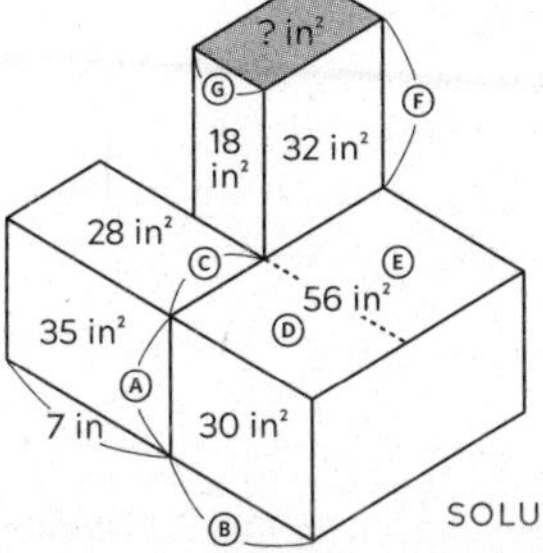

PUZZLE 196

Add the given areas: 27 + 33 = 60 in.²
Find the area with stripes: 60 − (5 × 8) = 20 in.²
Since 60 is triple 20, 27 must be triple Ⓐ:
 Ⓐ = 9 in.²
Ⓑ . . . 27 − 9 = 18 in.²
Area ⑦ is (5 × 7) − Ⓑ = 17 in.²

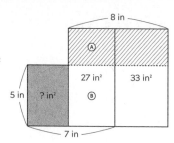

PUZZLE 197

Add the areas on top: 13 + 15 + 17 = 45 in.²
This is triple 15, so Ⓐ must be triple 4: Ⓐ = 12 in.
Add the other given areas: 11 + 19 = 30 in.²
Find their combined width: 12 − 3 = 9 in.
This is triple 3, so 30 must be triple ⑦.
Area ⑦ is 10 in.²

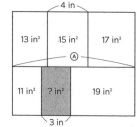

PUZZLE 198

Ⓐ . . . 12 ÷ 3 = 4 in.
Ⓑ . . . 36 ÷ 4 = 9 in.
Ⓒ . . . (3 × 9) − 10 = 17 in.²
This equals the given area of 17, so Ⓓ
 must equal 3.
Ⓔ . . . 9 ÷ 3 = 3 in.
The areas of ⑦ and 17 share a side, and
 their other sides, Ⓓ and Ⓔ, are equal.
Area ⑦ is 17 in.²

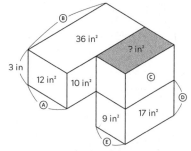

PUZZLE 199

Ⓐ . . . (6 × 7) − 19 = 23 in.²
Ⓑ . . . (6 × 8) − 23 = 25 in.²
Ⓒ . . . 50 − 25 = 25 in.²
This equals Ⓑ, so ⑦ must equal Ⓐ.
Area ⑦ is 23 in.²

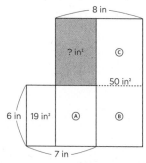

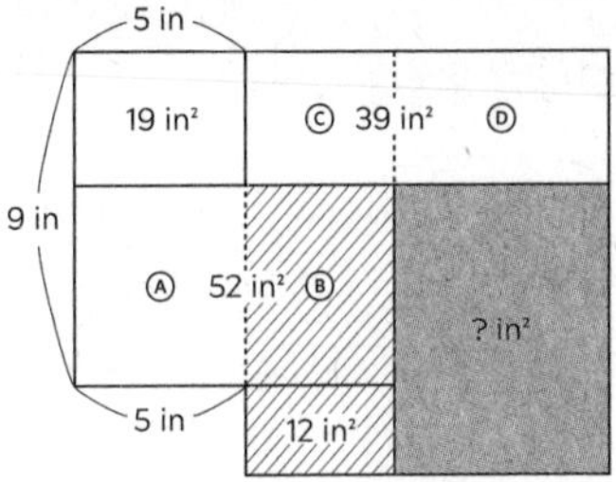

PUZZLE 200

Ⓐ . . . (9 × 5) – 19 = 26 in.^{2}

Ⓑ . . . 52 – 26 = 26 in.^{2}

This equals Ⓐ, so Ⓒ must equal 19.

Ⓓ . . . 39 – 19 = 20 in.^{2}

Add the areas with stripes: 26 + 12 = 38 in.^{2}

This is double Ⓒ, so (?) must be double Ⓓ.

Area (?) is 40 in.^{2}

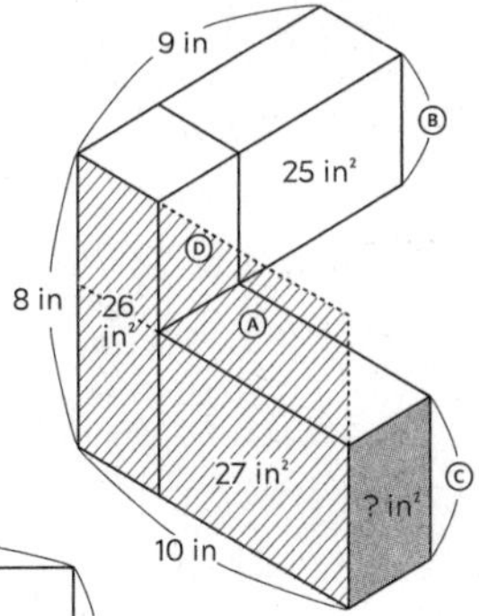

PUZZLE 201

Find the area with stripes: 8 × 10 = 80 in.^{2}

Ⓐ . . . 80 – 26 – 27 = 27 in.^{2}

This equals the given area of 27, so Ⓑ must equal Ⓒ.

Ⓑ + Ⓒ = 8, so Ⓑ and Ⓒ equal 4.

Ⓓ . . . (4 × 9) – 25 = 11 in.^{2}

Ⓓ and (?) share a side, and their other sides, Ⓑ and Ⓒ, are equal.

Therefore (?) must equal Ⓓ.

Area (?) is 11 in.^{2}

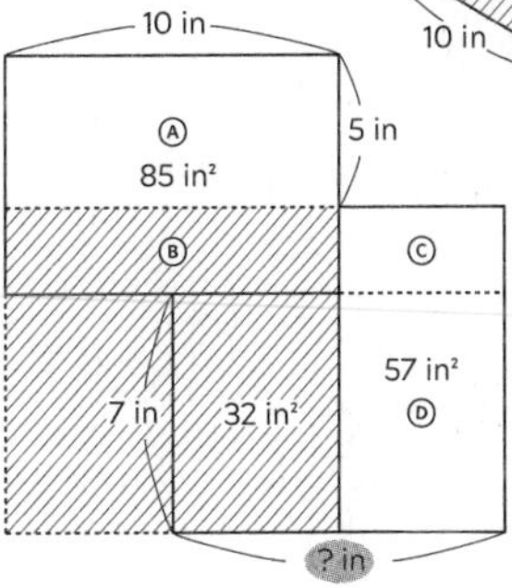

PUZZLE 202

Ⓐ . . . 5 × 10 = 50 in.^{2}

Ⓑ . . . 85 – 50 = 35 in.^{2}

Find the area with stripes:
(7 × 10) + 35 = 105 in.^{2}

This is triple Ⓑ, so 57 must be triple Ⓒ: Ⓒ = 19 in.^{2}

Ⓓ . . . 57 – 19 = 38 in.^{2}

Length (?) is (32 + Ⓓ) ÷ 7 = 10 in.

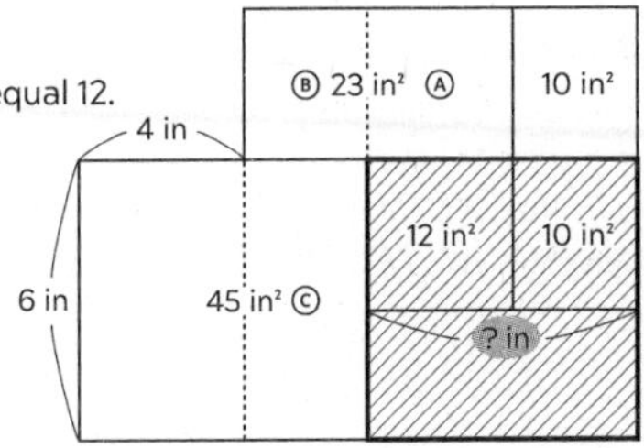

PUZZLE 203

The given areas of 10 are equal, so Ⓐ must equal 12.

Ⓑ . . . 23 – 12 = 11 in.^{2}

Add Ⓐ to the top area of 10: 12 + 10 = 22 in.^{2}

This is double Ⓑ, so the area with stripes must be double Ⓒ.

Ⓒ . . . 45 – (6 × 4) = 21 in.^{2}

Find the area with stripes: 21 × 2 = 42 in.^{2}

Length (?) is 42 ÷ 6 = 7 in.

PUZZLE 204

Ⓐ . . . 20 ÷ 4 = 5 in.

Ⓑ . . . 45 ÷ 5 = 9 in.

Ⓒ . . . (4 × 9) − 15 = 21 in.²

Ⓒ equals the given area of 21, and they share a side. Therefore Ⓓ must equal 4.

Ⓔ . . . 36 ÷ 4 = 9 in.

Ⓕ . . . 9 − 5 = 4 in.

Ⓒ and ⑦ share a side, and their other sides both equal 4.

Therefore ⑦ must equal Ⓒ.

Area ⑦ is 21 in.²

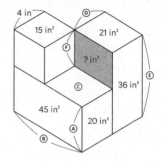

PUZZLE 205

Add the areas with width Ⓐ: 57 + 39 = 96 in.²

Ⓐ . . . 15 − 5 = 10 in.

This is double 5, so 96 must be double the area with stripes.

Find the area with stripes: 96 ÷ 2 = 48 in.²

Ⓑ . . . 48 − 24 = 24 in.²

Ⓒ . . . 40 − 24 = 16 in.²

Since 48 is triple Ⓒ, 57 must be triple ⑦.

Area ⑦ is 19 in.²

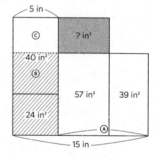

PUZZLE 206

Ⓐ . . . 7 × 5 = 35 in.²

Ⓑ . . . 49 − 35 = 14 in.²

This equals the given area of 14, so Ⓒ must equal 5.

Ⓓ . . . 15 − 5 − 5 = 5 in.

Therefore Ⓔ must also equal 14.

Ⓕ . . . 24 − 14 = 10 in.²

Ⓖ . . . 10 ÷ 5 = 2 in.

Length ⑦ is 7 − Ⓖ = 5 in.

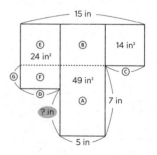

PUZZLE 207

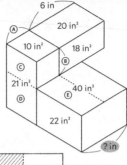

Since 20 is double 10, 6 must be double Ⓐ: Ⓐ = 3 in.

Ⓑ . . . 18 ÷ 6 = 3 in.

The areas of 10 and Ⓒ share a side, and Ⓐ equals Ⓑ.
Therefore Ⓒ must equal 10.

Ⓓ . . . 21 − 10 = 11 in.²

This is half of 22, so 10 must be half of Ⓔ: Ⓔ = 20 in.²

This is half of 40, so Ⓐ must be half of Ⓩ.

Length Ⓩ is 6 in.

PUZZLE 208

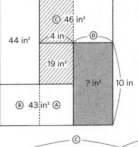

Ⓐ . . . (10 × 4) − 19 = 21 in.²

Ⓑ . . . 43 − 21 = 22 in.²

This is half of 44, so Ⓐ must be half the
area with stripes.

Find the area with stripes: 21 × 2 = 42 in.²

Ⓒ . . . 42 − 19 = 23 in.²

This is half of 46, so Ⓓ must equal 4.

Area Ⓩ is 10 × Ⓓ = 40 in.²

PUZZLE 209

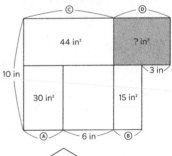

Since 30 is double 15, Ⓐ must be double Ⓑ.

Since 6 is also double 3, Ⓐ + 6 must be
double (Ⓑ + 3).

Ⓒ is double Ⓓ, so 44 must be double Ⓩ.

Area Ⓩ is 22 in.²

PUZZLE 210

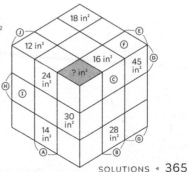

Since 14 is half of 28, Ⓐ must be half of Ⓑ.

Therefore 24 must be half of Ⓒ: Ⓒ = 48 in.²

This is triple 16, so Ⓓ must be triple Ⓔ.

Therefore 45 must be triple Ⓕ: Ⓕ = 15 in.²

This is half of 30, so Ⓖ must be half of Ⓗ.

Therefore Ⓘ must be double 18:
Ⓘ = 36 in.²

This is triple 12, so Ⓗ must be triple Ⓙ.

Therefore 30 must be triple Ⓩ.

Area Ⓩ is 10 in.²

PUZZLE 211

Ⓐ . . . (9 × 4) − 13 = 23 in.²
Add the areas with stripes: 17 + 29 = 46 in.²
Find their combined width: 12 − 4 = 8 in.
This is double 4, so 46 must be double Ⓑ: Ⓑ = 23 in.²
This equals Ⓐ, so ⑦ must equal 17.
Area ⑦ is 17 in.²

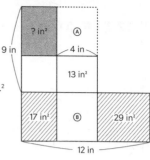

PUZZLE 212

Create area Ⓐ equal to the given area of 23.
They must have equal width, so Ⓑ must
    equal 7.
Ⓒ . . . 9 − 7 = 2 in.
Since 8 is quadruple Ⓒ, the area with
    stripes must be quadruple Ⓓ.
Ⓓ . . . 36 − 23 = 13 in.²
Find the area with stripes: 13 × 4 = 52 in.²
Area ⑦ is 52 − 23 = 29 in.²

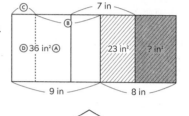

PUZZLE 213

Ⓐ . . . 3 × 4 = 12 in.²
Ⓑ . . . 23 − 12 = 11 in.²
This is half of 22, so 4 must be half of Ⓒ: Ⓒ = 8 in.
Find the area with stripes: (3 × 8) + 26 = 50 in.²
This is double 25, so Ⓒ must be double ⑦.
Length ⑦ is 4 in.

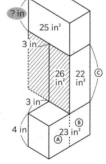

PUZZLE 214

Ⓐ . . . 65 ÷ (6 + 7) = 5 in.
Ⓑ . . . 8 − 5 = 3 in.
Ⓒ . . . 6 × 3 = 18 in.²
Ⓓ . . . 29 − 18 = 11 in.²
Ⓔ . . . 7 × 3 − 11 = 10 in.²
Ⓕ . . . 30 − 10 = 20 in.²
This is double Ⓔ, so ⑦ must be double Ⓓ.
Area ⑦ is 22 in.²

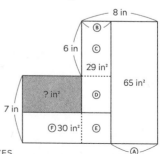

PUZZLE 215

(A) . . . (3 × 8) – 14 = 10 in.²

This equals the given area of 10,
   so (B) must equal 14.

And, (C) must equal 11.

(D) . . . 18 – 14 = 4 in.²

Add the areas above and below (D): 14 + 14 = 28 in.²

Add the areas with stripes: 10 + 10 + 11 + 11 = 42 in.²

Since 28 is septuple (D), 42 must be septuple (?).

Area (?) is 6 in.²

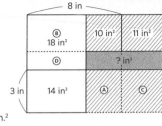

PUZZLE 216

(A) . . . (5 × 10) – 26 = 24 in.²

This is double 12, so (B) must be double (C).

Therefore 26 must be double (D): (D) = 13 in.²

(E) . . . 39 – 13 = 26 in.²

This equals the given area of 26, so (F) must equal 5.

Area (?) is 5 × (F) = 25 in.²

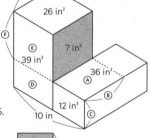

PUZZLE 217

(A) . . . 7 × 4 = 28 in.²

(B) . . . 58 – 28 = 30 in.²

Add the areas with stripes: 28 + 30 = 58 in.²

This is double 29, so (B) must be double (C): (C) = 15 in.²

(B) and (D) are the same height and width: (D) = 30 in.²

Area (?) is (C) + (D) = 45 in.²

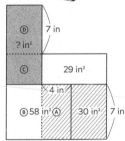

PUZZLE 218

(A) . . . (11 × 4) – 14 = 30 in.²

(B) . . . 45 – 30 = 15 in.²

This is half of (A), so (C) must be half of 4:
   (C) = 2 in.

(D) . . . 10 – 4 – 2 = 4 in.

This equals the given width of 4, so the
   area with stripes must equal (A).

Area (?) is (A) – (3 × (D)) = 18 in.²

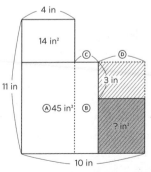

PUZZLE 219

Ⓐ ... 5 × 4 = 20 in.²

Ⓑ ... 33 – 20 = 13 in.²

This equals the given area of 13, so Ⓒ must equal 5.

Ⓓ ... 20 ÷ 5 = 4 in.

Ⓔ ... 14 – 4 – 5 = 5 in.

This equals the given width of 5, so the area with stripes must equal 33.

Ⓕ ... 33 – 13 = 20 in.²

This equals the adjacent area of 20, so Ⓖ must equal Ⓔ: Ⓖ = 5 in.

Area ⑦ is (Ⓓ + 5) × Ⓖ = 45 in.²

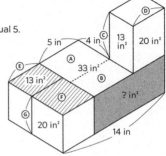

PUZZLE 220

Ⓐ ... 14 – 5 = 9 in.

Find the total area with stripes: (6 × 14) – 28 = 56 in.²

This is double 28, so 34 must be double Ⓑ: Ⓑ = 17 in.²

This equals the given area of 17, so Ⓒ must equal 6.

And, ⑦ must equal the striped area that shares its width.

Find this striped area: (6 × 9) – 28 = 26 in.²

Area ⑦ is 26 in.²

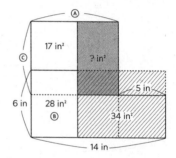

PUZZLE 221

Since 10 is double 5, the area with stripes must be double 48.

Add the given areas with width Ⓐ: 41 + 11 = 52 in.²

Add the given areas with width Ⓑ: 11 + 15 = 26 in.²

This is half of 52, so Ⓑ must be half of Ⓐ.

And, ⑦ must be half the striped area that shares its height.

Find this striped area: (48 × 2) – 41 – 11 = 44 in.²

Area ⑦ is 22 in.²

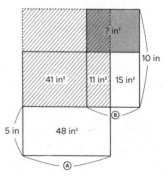

PUZZLE 222

Ⓐ ... 36 ÷ 4 = 9 in.

This equals the given length of 9,
  so Ⓑ must equal Ⓒ.

The given areas of 30 share a side,
  so Ⓓ must also equal Ⓒ.

Ⓑ × Ⓓ = 25, so Ⓑ, Ⓒ, and Ⓓ must all equal 5.

Ⓔ ... 30 ÷ 5 = 6 in.

Area ⑦ is (9 − Ⓒ) × Ⓔ = 24 in.²

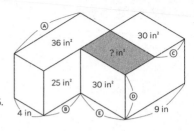

PUZZLE 223

Find the area with stripes: (7 × 9) − 42 = 21 in.²

This is half of 42, so 13 must be half of Ⓐ: Ⓐ = 26 in.²

This equals the given area of 26, so Ⓑ must equal 9.

Add the given areas of width 9: 42 + 57 = 99 in.²

Find their combined height:

Length ⑦ is 99 ÷ 9 = 11 in.

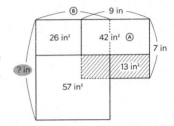

PUZZLE 224

Add the given areas of width 7: 24 + 25 = 49 in.²

Find their combined height: 49 ÷ 7 = 7 in.

Ⓐ ... 10 − 7 = 3 in.

Find the area with stripes: 3 × 7 = 21 in.²

Ⓑ ... 21 − 10 = 11 in.²

Since 6 is double Ⓐ, ⑦ must be double Ⓑ.

Area ⑦ is 22 in.²

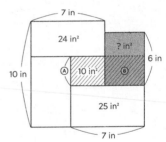

PUZZLE 225

Ⓐ ... 6 × 5 = 30 in.²

Ⓑ ... 52 − 30 = 22 in.²

Ⓑ and Ⓒ share a side, and their other
  sides both equal 6: Ⓒ = 22 in.²

Ⓓ ... 4 × 6 = 24 in.²

Add Ⓒ and Ⓓ: 22 + 24 = 46 in.²

This is double 23, so 6 must be double Ⓔ: Ⓔ = 3 in.

Ⓕ ... 18 ÷ 3 = 6 in.

⑦ and Ⓒ share a side, and their other sides both equal 6:

Area ⑦ is 22 in.²

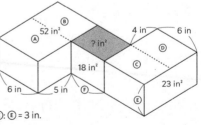

PUZZLE 226

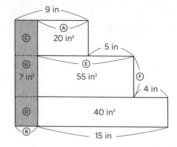

Ⓐ ... 15 − 5 − 4 = 6 in.

Ⓑ ... 9 − 6 = 3 in.

This is half of Ⓐ, so Ⓒ must be half of 20: Ⓒ = 10 in.²

Since 15 is quintuple Ⓑ, 40 must be quintuple Ⓓ: Ⓓ = 8 in.²

Ⓔ ... 15 − 4 = 11 in.

Ⓕ ... 55 ÷ 11 = 5 in.

Ⓖ ... 5 × 3 = 15 in.²

Area ⑦ is Ⓒ + Ⓖ + Ⓓ = 33 in.²

PUZZLE 227

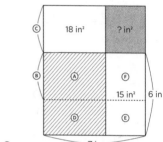

Find the area with stripes: (6 × 7) − 15 = 27 in.²

Draw Ⓐ equal to the given area of 18: Ⓑ must equal Ⓒ.

Ⓓ ... 27 − 18 = 9 in.²

Since 27 is triple 9, 15 must be triple Ⓔ: Ⓔ = 5 in.²

Ⓕ ... 15 − 5 = 10 in.²

From Ⓑ = Ⓒ, ⑦ must equal Ⓕ.

Area ⑦ is 10 in.²

PUZZLE 228

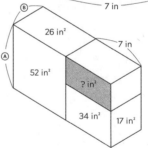

Since 52 is double 26, Ⓐ must be double Ⓑ.

Since 34 is double 17, 7 must also be double Ⓑ.

Therefore Ⓐ equals 7.

Area ⑦ is (Ⓐ × 7) − 34 = 15 in.²

PUZZLE 229

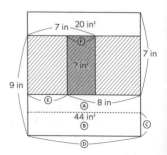

Draw Ⓐ equal to the given area of 20.

Ⓑ ... 44 − 20 = 24 in.²

The areas of Ⓐ and 20 must have equal height, so:

Ⓒ ... 9 − 7 = 2 in.

Ⓓ ... 24 ÷ 2 = 12 in.

Find the area with stripes: (7 × 12) − 20 = 64 in.²

Ⓔ ... 12 − 8 = 4 in.

Ⓕ ... 7 − 4 = 3 in.

Ⓓ is quadruple Ⓕ, so 64 must be quadruple ⑦.

Area ⑦ is 16 in.²

PUZZLE 230

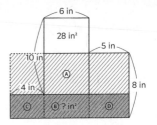

Ⓐ . . . (10 × 6) − 28 = 32 in.²

Ⓑ . . . (8 × 6) − 32 = 16 in.²

Add Ⓐ and Ⓑ: 32 + 16 = 48 in.²

This is triple Ⓑ, so the area with stripes must be triple ⑦.

Find the area with stripes: 8 × (4 + 6 + 5) = 120 in.²

Area ⑦ is 120 ÷ 3 = 40 in.²

PUZZLE 231

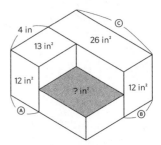

The given areas of 12 are the same height, so Ⓐ must equal Ⓑ.

Therefore, since 26 is double 13, Ⓒ must be double 4: Ⓒ = 8 in.

Area ⑦ is (4 × Ⓒ) − 13 = 19 in.²

PUZZLE 232

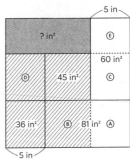

The areas of Ⓐ and 36 are the same height and width: Ⓐ = 36 in.²

Ⓑ . . . 81 − 36 = 45 in.²

This equals the given area of 45, so Ⓒ and Ⓓ must equal 36.

Ⓔ . . . 60 − 36 = 24 in.²

Add Ⓐ and Ⓒ: 36 + 36 = 72 in.²

This is triple Ⓔ, so the area with stripes must be triple ⑦.

Add the areas with stripes: 36 + 45 + 36 + 45 = 162 in.²

Area ⑦ is 162 ÷ 3 = 54 in.²

PUZZLE 233

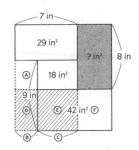

Ⓐ . . . (8 × 7) − 29 − 18 = 9 in.²

This is half of 18, so Ⓑ must be half of Ⓒ.

Find the area with stripes: (9 × 7) − 9 − 18 = 36 in.²

Since Ⓑ is half of Ⓒ, 36 must be triple Ⓓ: Ⓓ = 12 in.²

And, Ⓔ must be double Ⓓ: Ⓔ = 24 in.²

Ⓕ . . . 42 − 24 = 18 in.²

This is half the area with stripes, so ⑦ must be half the total area to its left.

Area ⑦ is (8 × 7) ÷ 2 = 28 in.²

PUZZLE 234

The given areas of 14 must be the same width.
Draw Ⓐ equal to the given area of 21.
It must also be the same width, so Ⓑ must equal 7.
Ⓒ . . . 10 − 7 = 3 in.
Ⓓ . . . 48 − 21 = 27 in.²
Ⓔ . . . 27 ÷ 3 = 9 in.
Find the area with stripes: (9 × 7) − 21 = 42 in.²
This is triple 14, so Ⓔ must be triple Ⓕ: Ⓕ = 3 in.
Length ⑦ is 9 − Ⓕ = 6 in.

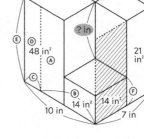

PUZZLE 235

Ⓐ . . . 6 × 5 = 30 in.²
Ⓑ . . . 59 − 30 = 29 in.²
This equals the given area of 29, so Ⓒ
must equal 41.
Ⓓ . . . 64 − 41 = 23 in.²
This equals the given area of 23, so the
area with stripes must equal Ⓒ.
Ⓔ . . . 41 − 29 = 12 in.²
Ⓕ . . . 43 − 12 = 31 in.²
This equals the given area of 31, so the
area with stripes must equal 23 + ⑦.
Area ⑦ is 41 − 23 = 18 in.²

PUZZLE 236

Add these two areas: 12 + 26 = 38 in.²
This is double 19, so 26 must be double Ⓐ:
Ⓐ = 13 in.²
Add the lower areas with stripes:
13 + 26 + 39 = 78 in.²
This is triple 26, so the upper area with
stripes must be triple 11.
Find this area: 11 × 3 = 33 in.²
Length ⑦ is 33 ÷ 3 = 11 in.

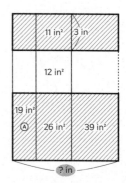

PUZZLE 237

(A) . . . 9 ÷ 3 = 3 in.

(B) . . . 15 ÷ 3 = 5 in.

(C) . . . 20 ÷ 5 = 4 in.

(D) . . . 3 × 4 = 12 in.²

(E) . . . 27 − 12 = 15 in.²

(F) . . . 3 + 5 = 8 in.

This is double (C), so the area with stripes must be double (E): 30 in.²

(G) . . . 30 − 20 = 10 in.²

This is half the adjacent area of 20, so 12 must be half of (?).

Area (?) is 24 in.²

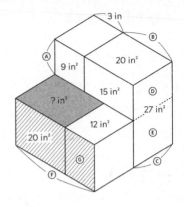

PUZZLE 238

(A) . . . (4 × 8) − 17 = 15 in.²

(B) . . . 4 × 7 = 28 in.²

Add the areas with stripes: 28 + 17 = 45 in.²

This is triple (A), so (C) must be triple 18: (C) = 54 in.²

(D) . . . 99 − 54 = 45 in.²

This equals the area with stripes, so:

Length (?) is 4 in.

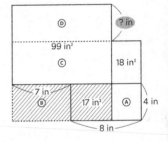

PUZZLE 239

Add the given areas on left: 9 + 40 + 8 = 57 in.²

Add the given areas on right: 20 + 40 = 60 in.²

This is triple 20, so 57 must be triple the area with stripes.

Find this area: 57 ÷ 3 = 19 in.²

(A) . . . 19 − 9 = 10 in.²

(B) . . . 40 − 10 = 30 in.²

This is triple (A), so (?) must be triple 13.

Area (?) is 39 in.²

PUZZLE 240

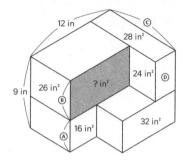

Ⓐ . . . (16 + 32) ÷ 12 = 4 in.

Ⓑ . . . 9 – 4 = 5 in.

Ⓒ . . . (26 + 24) ÷ 5 = 10 in.

This is double Ⓑ, so 28 must be double Ⓓ:
 Ⓓ = 14 in.²

Area ⑦ is (12 × Ⓑ) – Ⓓ = 46 in.²

PUZZLE 241

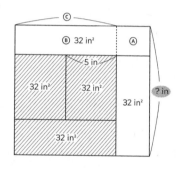

Find the area with stripes: 32 × 3 = 96 in.²

This is triple the area of 32 below Ⓐ, so Ⓑ must be triple Ⓐ.

Therefore the top area of 32 is quadruple Ⓐ:
 Ⓐ = 8 in.²

Ⓑ . . . 32 – 8 = 24 in.²

The areas of 32 with equal height must have equal width, so Ⓒ must be double 5:
 Ⓒ = 10 in.

Add Ⓑ to the area with stripes: 24 + 96 = 120 in.²

Length ⑦ is 120 ÷ Ⓒ = 12 in.

PUZZLE 242

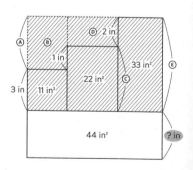

Ⓐ . . . 2 + 1 = 3 in.

This equals the given height of 3, so Ⓑ must equal 11.

Ⓒ . . . 1 + 3 = 4 in.

This is double 2, so 22 must be double Ⓓ:
 Ⓓ = 11 in.²

Add the areas with stripes:
 11 + 11 + 11 + 22 + 33 = 88 in.²

This is double 44, so Ⓔ must be double ⑦.

Ⓔ . . . 2 + 1 + 3 = 6 in.

Length ⑦ is 3 in.

PUZZLE 243

Draw another box of equal height.
Find the area with stripes: 14 × 2 = 28 in.²
This is half of 56, so Ⓐ must be half of 10: Ⓐ = 5 in.
Area ⑦ is (10 × Ⓐ) ÷ 2 = 25 in.²

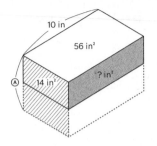

PUZZLE 244

Since 8 is double 4, Ⓐ must be double 23:
  Ⓐ = 46 in.²
Ⓑ . . . (8 × 10) − 46 = 34 in.²
Find the total area of width 9: 8 × 9 = 72 in.²
Area ⑦ is 72 − Ⓑ − 20 = 18 in.²

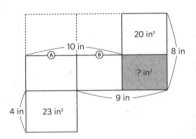

PUZZLE 245

Since 20 is double 10, and 6 is double 3, the
  areas with stripes must have equal height.
Therefore Ⓐ must equal Ⓑ.
Ⓒ . . . 3 + 1 = 4 in.
Ⓓ . . . 6 − 2 = 4 in.
This equals Ⓒ, so the areas of 30 and
  ⑦ are the same height and width.
Area ⑦ is 30 in.²

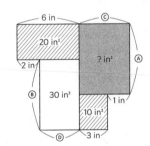

PUZZLE 246

Ⓐ × Ⓑ = 28, and Ⓒ × Ⓓ = 30.
Therefore Ⓐ × Ⓑ × Ⓒ × Ⓓ = 28 × 30 = 840.
Ⓐ × Ⓓ = 24.
Therefore 840 ÷ 24 = Ⓑ × Ⓒ = 35.
Ⓑ × Ⓒ = ⑦, therefore:
Area ⑦ is 35 in.²

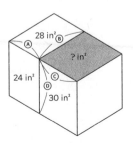

PUZZLE 247

Ⓐ ... 9 − 6 = 3 in.

Since 9 is triple Ⓐ, 51 must be triple Ⓑ: Ⓑ = 17 in.²

Ⓒ ... 51 − 17 = 34 in.²

This equals the given area of 34: Ⓓ = 6 in.

Therefore Ⓔ must equal 27.

Ⓕ ... 45 − 27 = 18 in.²

Add the areas with stripes: 27 + 27 = 54 in.²

This is triple Ⓕ, so Ⓖ must be triple Ⓕ.

Ⓖ ... 6 + 6 = 12 in.

Length ? is 4 in.

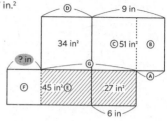

PUZZLE 248

Ⓐ ... 4 × 5 = 20 in.²

Ⓑ ... 41 − 20 = 21 in.²

This equals the given area of 21, so Ⓒ
   must equal 4, and Ⓓ must equal 23.

Ⓔ ... 43 − 23 = 20 in.²

This equals the given area of 20: Ⓕ = 31 in.²

Ⓖ ... 51 − 31 = 20 in.²

This equals the given area of 20: Ⓗ = Ⓘ.

Find the lower area with stripes: 23 + 23 = 46 in.²

Find the upper area with stripes: 20 + 26 = 46 in.²

These equal areas are the same width, so they must be the same height.

Length ? is 4 + Ⓒ = 8 in.

PUZZLE 249

Ⓐ ... 35 ÷ 5 = 7 in.

Ⓑ ... 63 ÷ 7 = 9 in.

Ⓒ ... 9 × 5 = 45 in.²

Ⓓ ... 57 − 45 = 12 in.²

Find the area with stripes:
   12 + (9 × 4) = 48 in.²

This is quadruple 12, so 64 must be quadruple Ⓔ: Ⓔ = 16 in.²

Ⓕ ... 64 − 16 = 48 in.²

Ⓖ ... 48 ÷ 4 = 12 in.

Area ? is Ⓐ × Ⓖ = 84 in.²

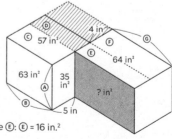

PUZZLE 250

(A) ... 12 − 7 = 5 in.

(B) ... 40 ÷ 5 = 8 in.

Find the area with stripes: (8 × 7) − 28 = 28 in.^{2}

This equals the given area of 28, so 40 must be double (C): (C) = 20 in.^{2}

(D) ... 28 − 18 = 10 in.^{2}

This is half of (C), so (E) must be half of (?).

(E) ... 21 − 10 = 11 in.^{2}

Area (?) is 22 in.^{2}

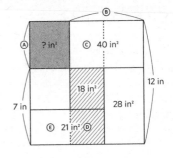

PUZZLE 251

The given lengths of 10 are equal: (A) = (B).

Therefore (C) = 15 in.^{2}

(D) ... 28 − 15 = 13 in.^{2}

This equals the given area of 13: (E) = (C).

(E) equals the given area of 15, so its width is (A).

Therefore (F) = 14 + 13 = 27 in.^{2}

(G) ... 25 − 15 = 10 in.^{2}

Add the areas with stripes: 15 + 15 = 30 in.^{2}

This is triple (G), so (F) + 14 + 13 must be triple (H): (H) = 18 in.^{2}

Area (?) is (F) + (H) = 45 in.^{2}

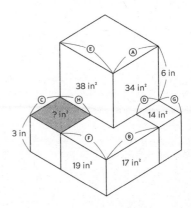

PUZZLE 252

Since 34 is double 17, and 6 is double 3: (A) = (B).

Therefore (C) = (D).

Similarly, since 38 is double 19: (E) = (F).

Therefore (G) = (H).

The given area of 14 has sides (D) and (G).

These equal sides (C) and (H) of area (?).

Area (?) is 14 in.^{2}

PUZZLE 253

Ⓐ . . . (10 × 4) − 29 = 11 in.²

Ⓑ . . . 25 − 11 = 14 in.²

Since 8 is double 4, Ⓒ is double Ⓐ:
Ⓒ = 22 in.²

Add Ⓑ and Ⓒ: 14 + 22 = 36 in.²

This equals the given area of 36:
Ⓓ = Ⓒ.

Area ⑦ is (9 × 8) − Ⓓ = 50 in.²

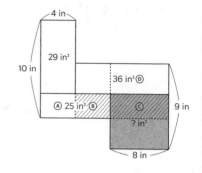

PUZZLE 254

Ⓐ . . . (5 × 8) − 23 = 17 in.²

This equals the given area of 17:
Ⓑ = 23 in.²

Therefore Ⓒ = Ⓓ.

Since 62 is double 31 with equal height:
Ⓔ is double Ⓕ.

Therefore the area with stripes is
double Ⓑ: 46 in.²

Area ⑦ is 46 − 17 = 29 in.²

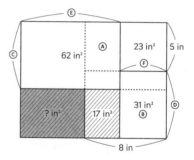

PUZZLE 255

The given areas of 51 have equal height:
Ⓐ = Ⓑ.

Therefore the area of 52 is a square.

Similarly, the area of 13 is a square.

Divide 52 into four smaller squares:
Ⓒ = 52 ÷ 4 = 13 in.²

This equals the given area of 13. Both
are squares, so Ⓓ = Ⓔ.

Therefore Ⓐ is double Ⓔ.

Ⓔ × Ⓔ = 13, and Ⓐ × Ⓔ = ⑦. Therefore:

Area ⑦ is 13 × 2 = 26 in.²

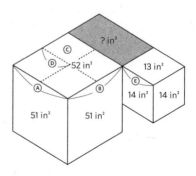

PUZZLE 256

(A) . . . 49 ÷ (4 + 3) = 7 in.

(B) . . . 4 × 7 = 28 in.²

This is double 14, so (C) must be double the area with stripes.

(D) . . . 13 − 4 − 3 = 6 in.

(C) . . . 6 × 7 = 42 in.²

Therefore the area with stripes = 21 in.²

(E) . . . 21 − (45 − 42) = 18 in.²

(F) . . . 18 ÷ 6 = 3 in.

(G) . . . 33 − 18 = 15 in.²

Length (?) is (G) ÷ (F) = 5 in.

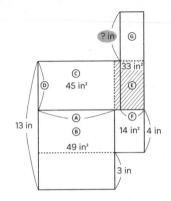

PUZZLE 257

Draw area (A) equal to 9: (B) = (C).

The areas of 27 have equal width: (D) = (E).

Therefore (F) = 19 in.²

(G) . . . 37 − 19 = 18 in.²

The area with stripes = 9 + 9 = 18 in.²

This equals (G), so (H) = (F).

Area (?) is (A) + (H) = 28 in.²

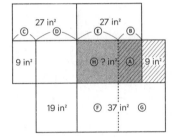

PUZZLE 258

Draw length (A) equal to 8.

The new area with stripes = (7 × 8) − 25 = 31 in.²

It shares a side with the given area of 31: (B) = 7 in.

The lower area with stripes = (7 × 9) − 31 = 32 in.²

It shares a side with the given area of 32: (B) = (?).

Length (?) is 7 in.

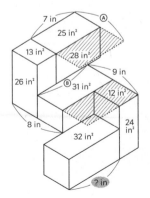

PUZZLE 259

The given areas of 41 are equal: Ⓐ = 29 in.^{2}
Ⓑ ... 66 − 29 = 37 in.^{2}
This equals the given area of 37: Ⓒ = 29 in.^{2}
Ⓓ ... 64 − 29 = 35 in.^{2}
From the given areas of 41: Ⓔ = Ⓓ.
Ⓕ ... 72 − 35 = 37 in.^{2}
Add the areas with stripes: 29 + 41 = 70 in.^{2}
This is double Ⓓ, so ⑦ must be double Ⓕ.
Area ⑦ is 74 in.^{2}

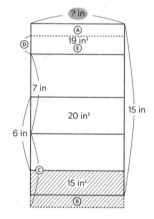

PUZZLE 260

Draw area Ⓐ equal to 5, then move it to Ⓑ.
Add the areas with stripes: 15 + 5 = 20 in.^{2}
This equals the given area of 20: Ⓒ = 6 in.
After moving Ⓐ, the total height remains 15.
Ⓓ ... 15 − 7 − 6 = 2 in.
Ⓔ ... 19 − 5 = 14 in.^{2}
Length ⑦ is Ⓔ ÷ Ⓓ = 7 in.

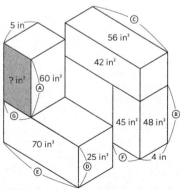

PUZZLE 261

Ⓐ ... 60 ÷ 5 = 12 in.
Ⓑ ... 48 ÷ 4 = 12 in.
Ⓒ ... 56 ÷ 4 = 14 in.
Ⓓ ... 25 ÷ 5 = 5 in.
Ⓔ ... 70 ÷ 5 = 14 in.
This equals Ⓒ, so Ⓕ = Ⓖ.
The given area of 45 has sides of Ⓑ and Ⓕ.
These equal sides Ⓐ and Ⓖ of area ⑦.
Area ⑦ is 45 in.^{2}

PUZZLE 262

Since 38 is double 19, and 6 is double 3:
    Ⓐ = Ⓑ.
Therefore Ⓒ = 42 in.²
Ⓓ . . . 70 − 42 = 28 in.²
Since Ⓐ equals Ⓑ, Ⓔ = Ⓓ.
Ⓕ . . . 38 − 28 = 10 in.²
Since 6 is double 3, Ⓕ is double Ⓖ:
    Ⓖ = 5 in.²
Ⓗ . . . 19 − 5 = 14 in.²
Ⓒ is triple Ⓗ, so ⑦ is triple Ⓖ.
Area ⑦ is 15 in.²

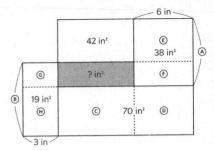

PUZZLE 263

The given heights of 8 are equal: Ⓐ = Ⓑ.
Draw width Ⓒ equal to Ⓓ: The area with
    stripes = 19 in.²
From Ⓒ = Ⓓ, 7 + Ⓔ must equal 10:
    Ⓔ = 3 in.
Ⓕ . . . 31 − 19 = 12 in.²
Ⓐ . . . 12 ÷ 3 = 4 in.
Ⓖ . . . 8 − 4 = 4 in.
This equals Ⓐ, so Ⓗ = 31 in.²
Ⓘ . . . 4 × 7 = 28 in.²
Area ⑦ is Ⓗ + Ⓘ = 59 in.²

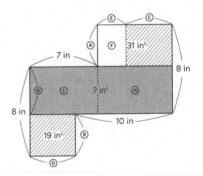

PUZZLE 264

Ⓐ . . . 32 ÷ 4 = 8 in.
This is double 4, so the area with stripes
    must be double 23: 46 in.²
Since 26 is double 13, Ⓑ must be double 17:
    Ⓑ = 34 in.²
This equals the given area of 34: Ⓒ = Ⓓ.
Therefore the area with stripes + 26
    must equal 34 + ⑦.
Area ⑦ is 46 + 26 − 34 = 38 in.²

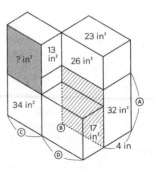

PUZZLE 265

Ⓐ ... (4 × 10) − 25 = 15 in.²
Ⓑ ... 36 − 15 = 21 in.²
Ⓒ ... (10 × 5) − 22 = 28 in.²
Ⓓ ... 49 − 28 = 21 in.²
This equals Ⓑ, so 56 is double Ⓔ: Ⓔ = 28 in.²
Ⓔ = Ⓒ, so they must have the same width.
Length ⑦ is 5 in.

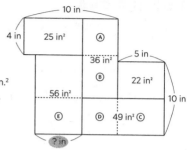

PUZZLE 266

Since 58 is double 29,
    ⑦ must be triple Ⓐ.
Since 65 is quintuple 13,
    ⑦ must be sextuple Ⓑ.
Therefore Ⓐ must be double Ⓑ.
Ⓐ + 5 + Ⓑ = ⑦.
Therefore 5 must be triple Ⓑ.
Therefore ⑦ must be double 5.
Length ⑦ is 10 in.

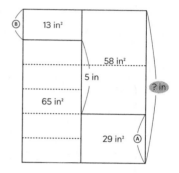

PUZZLE 267

Ⓐ ... 6 × 3 = 18 in.²
Ⓑ ... 35 − 18 = 17 in.²
Ⓒ ... 6 × 4 = 24 in.²
Add Ⓑ and Ⓒ: 17 + 24 = 41 in.²
This equals the given area of 41:
    Ⓓ = 6 in.
Ⓔ ... 6 × 2 = 12 in.²
Ⓕ ... 39 − 12 = 27 in.²
From Ⓓ = 6, Ⓖ = Ⓕ.
Ⓗ ... 6 × 5 = 30 in.²
Area ⑦ is Ⓗ + Ⓖ = 57 in.²

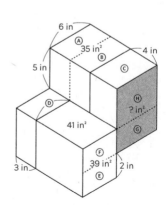

PUZZLE 268

Add Ⓐ and Ⓑ: 53 + 31 = 84 in.^{2}

Add Ⓒ and Ⓓ: 25 + 17 = 42 in.^{2}

This is half of 84, so 9 must be half of Ⓕ: Ⓕ = 18 in.

Add Ⓓ and Ⓔ: 17 + 11 = 28 in.^{2}

Since 84 is triple 28, Ⓕ must be triple Ⓖ: Ⓖ = 6 in.

Add Ⓑ and Ⓒ: 31 + 25 = 56 in.^{2}

This is double 28, so Ⓟ must be double Ⓖ.

Length Ⓟ is 12 in.

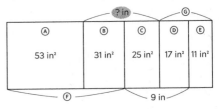

PUZZLE 269

The given heights of 3 are equal: Ⓐ = Ⓑ.

Therefore the area with stripes
= 19 + 23 + 27 = 69 in.^{2}

This is triple 23, so 3 × 14 must be triple Ⓒ:
Ⓒ = 14 in.^{2}

From the given heights of 3, Ⓟ = Ⓒ.

Area Ⓟ is 14 in.^{2}

PUZZLE 270

Since 51 is triple 17, Ⓐ is triple Ⓑ.

Therefore 20 + Ⓒ must be triple 15:
Ⓒ . . . (15 × 3) − 20 = 25 in.^{2}

Add Ⓒ to the given area of 15:
25 + 15 = 40 in.^{2}

This is double 20: Ⓓ must be double Ⓔ.

Therefore the area with stripes must be double 12: 24 in.^{2}

Length Ⓟ is (12 + 24 + 51 + 17) ÷ 13 = 8 in.

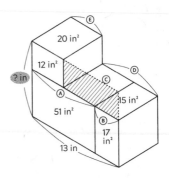

PUZZLE 271

Find the area with stripes: 8 × 7 = 56 in.²
This is double 28, so Ⓐ must be double
  17: Ⓐ = 34 in.²
Ⓑ . . . 3 × 7 = 21 in.²
Ⓒ . . . 56 − 21 = 35 in.²
Add Ⓐ and Ⓒ: 34 + 35 = 69 in.²
This is triple 23, so Ⓑ must be triple Ⓓ:
  Ⓓ = 7 in.²
Area ⑦ is Ⓑ + Ⓓ = 28 in.²

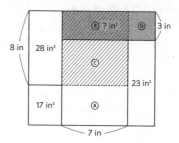

PUZZLE 272

On top, 52 is double 26.
Below, 13 + Ⓐ + 13 must be
  double (Ⓑ + 26).
Therefore Ⓐ = Ⓑ + Ⓑ + 26.
This is double (13 + Ⓑ).
Therefore ⑦ is double 4.
Length ⑦ is 8 in.

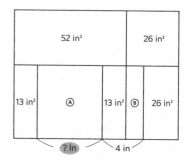

PUZZLE 273

Ⓐ . . . 40 ÷ 8 = 5 in.
Ⓑ . . . 15 ÷ 5 = 3 in.
Ⓒ . . . 24 ÷ 3 = 8 in.
This equals the given length of 8: Ⓓ = Ⓔ.
Ⓕ . . . 32 ÷ 8 = 4 in. This equals Ⓑ + 1.
Therefore, if Ⓖ = Ⓗ, Ⓘ = 1 in.
Ⓙ . . . 30 − 26 = 4 in.²
Ⓔ . . . 4 ÷ 1 = 4 in.
Ⓚ . . . 8 − 4 = 4 in. This equals Ⓔ.
⑦ and Ⓛ share a side, and their other sides are equal: ⑦ = Ⓛ.
Find the area with stripes: 4 × 4 = 16 in.²
Ⓛ . . . 30 − 16 = 14 in.²
Area ⑦ is 14 in.²

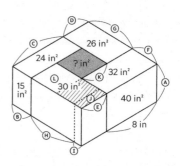

PUZZLE 274

Find the area with stripes: $7 \times 4 = 28$ in.^{2}

This is double 14, so $45 + ⑦$ must be double $(13 + ⑦)$.

Therefore $45 + ⑦$ equals $26 + ⑦ + ⑦$.

Therefore 45 equals $26 + ⑦$.

Area ⑦ is $45 - 26 = 19$ in.^{2}

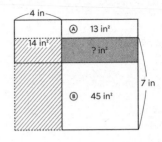

PUZZLE 275

Ⓐ ... $7 \times 5 = 35$ in.^{2}

Ⓑ ... $5 \times 5 = 25$ in.^{2}

Add the lower areas with stripes: $25 + 17 = 42$ in.^{2}

Draw Ⓒ equal to the given height of 7.

The new area with height Ⓒ must equal Ⓐ.

Ⓓ ... $42 - Ⓐ = 7$ in.^{2}

Ⓐ is quintuple Ⓓ, so the upper area with stripes must be quintuple Ⓔ.

Find this area: $17 + (7 \times 4) = 45$ in.^{2}

Therefore Ⓔ = 9 in.^{2}

Since 42 is sextuple Ⓓ, ⑦ must be sextuple Ⓔ.

Area ⑦ is 54 in.^{2}

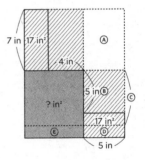

PUZZLE 276

Since 34 is double 17, Ⓐ must be double Ⓑ.

Ⓒ = $10 - Ⓐ$, and Ⓓ = $5 - Ⓑ$.

Since 10 is also double 5, Ⓒ must be double Ⓓ.

Therefore ⑦ must be double 26.

Area ⑦ is 52 in.^{2}

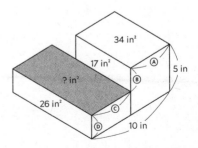

PUZZLE 277

Ⓐ . . . (20 + 35) ÷ 11 = 5 in.

Ⓑ . . . 35 ÷ 5 = 7 in.

Ⓒ . . . 11 − 7 = 4 in.

Find the area with stripes:
 (7 × 15) − 34 − 35 = 36 in.²

This equals the given area of 36: ⑦ = Ⓒ.

Length ⑦ is 4 in.

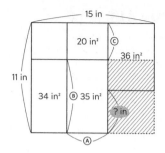

PUZZLE 278

Ⓐ . . . (3 × 8) − 16 = 8 in.² This is half of 16.

Therefore the area of the middle
 column = (16 + 56) ÷ 2 = 36 in.²

Consider the areas on left: 57 is triple 19.
 Therefore 36 must be quadruple the
 area with stripes.

Find this area: 36 ÷ 4 = 9 in.²

Ⓑ . . . 36 − 9 − 19 = 8 in.²

This equals Ⓐ, therefore:

Length ⑦ is 3 in.

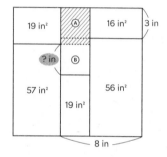

PUZZLE 279

Divide 52 into two equal areas: Ⓐ = 26 in.²

This equals the given area of 26: Ⓑ = Ⓒ.

Divide 38 into two equal areas: Ⓓ = 19 in.²

Draw length Ⓔ equal to Ⓕ.

Ⓓ and Ⓖ have equal sides: Ⓖ = 19 in.²

Ⓗ . . . 32 − 19 = 13 in.²

Ⓕ + 10 equals Ⓔ + ⑦ + 8,
 so from Ⓔ = Ⓕ: ⑦ = 2 in.

Since 8 is quadruple 2, Ⓙ must
 be quadruple Ⓗ: Ⓙ = 52 in.²

Add the areas with stripes:
 52 + 32 = 84 in.²

This is double 42, so 26
 must be double ⑦.

Area ⑦ is 13 in.²

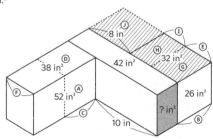

PUZZLE 280

Since 58 is double 29: Ⓐ is double Ⓑ.

Since 34 is also double 17, the area with stripes must be double Ⓒ.

Therefore 6 must be triple Ⓓ: Ⓓ = 2 in.

Length ⑦ is 6 − Ⓓ = 4 in.

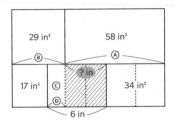

PUZZLE 281

Draw Ⓐ equal to the given height of 5.

Therefore Ⓑ must equal the given area of 17.

Ⓒ . . . 11 − 5 = 6 in.

Ⓓ . . . (6 × 9) − 17 − 11 = 26 in.^{2}

Ⓔ . . . 39 − 26 = 13 in.^{2}

This is half of Ⓓ, so Ⓕ must half of Ⓒ: Ⓕ = 3 in.

Length ⑦ is Ⓒ + Ⓕ = 9 in.

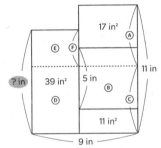

PUZZLE 282

Ⓐ . . . 27 ÷ 9 = 3 in.

Ⓑ . . . 12 ÷ 3 = 4 in.

Ⓒ equals Ⓑ: Ⓒ = 4 in.

Ⓓ . . . 20 ÷ 4 = 5 in.

Ⓔ . . . 45 ÷ 5 = 9 in.

This equals the given height of 9: Ⓕ = Ⓖ.

Draw length Ⓗ equal to Ⓘ: The area with stripes must be 32.

From Ⓗ = Ⓘ, Ⓐ + Ⓙ must equal Ⓓ: Ⓙ = 2 in.

Ⓚ . . . 40 − 32 = 8 in.^{2}

Ⓕ and Ⓖ . . . 8 ÷ 2 = 4 in.

Ⓗ and Ⓘ . . . 32 ÷ 4 = 8 in.

Ⓘ is double Ⓖ, so ⑦ must be double 15.

Area ⑦ is 30 in.^{2}

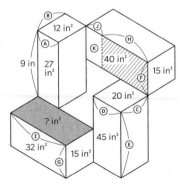

PUZZLE 283

Draw Ⓐ of equal width to the given area of 7.
Ⓑ . . . 22 − 7 = 15 in.²
Ⓒ . . . 11 − 5 = 6 in.
Draw Ⓓ of equal height to the given area of 19.
Ⓔ . . . 41 − 19 = 22 in.²
Ⓕ . . . 11 − 6 = 5 in.
Ⓖ . . . 5 × 6 = 30 in.²
This is double Ⓑ, so Ⓔ must be double Ⓒ.
Area ⑦ is 11 in.²

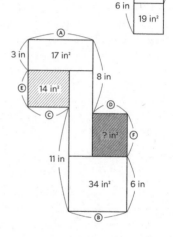

PUZZLE 284

Since 34 is double 17, and 6 is double 3:
Ⓐ = Ⓑ.
Therefore Ⓒ = Ⓓ.
The total height 3 + Ⓔ + 11 must
equal 8 + Ⓕ + 6.
Ⓔ + 14 equals Ⓕ + 14, so Ⓔ = Ⓕ.
The areas of ⑦ and 14 have equal sides.
Area ⑦ is 14 in.²

PUZZLE 285

Since 6 is double 3, Ⓐ + 16 + 19 must be
double 26.
Ⓐ . . . (26 × 2) − 16 − 19 = 17 in.²
Ⓑ . . . 51 − 17 = 34 in.²
This is double Ⓐ, so Ⓒ must be
double 16: Ⓒ = 32 in.²
And, area Ⓓ with stripes must be
half of 64: Ⓓ = 32 in.²
Add these areas facing right:
Ⓒ + 16 + 51 = 99 in.²
Since 12 is double 6, Ⓓ + 64 + ⑦
must be double 99.
Area ⑦ is (99 × 2) − 64 − Ⓓ = 102 in.²

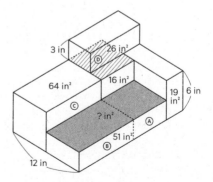

PUZZLE 286

Ⓐ . . . (45 ÷ 5) − 4 = 5 in.

Ⓑ . . . (42 ÷ 6) − 2 = 5 in.

The areas of ⑦ and 49 are the same width.

Ⓒ . . . 54 ÷ 9 = 6 in.

Add this to the given height of 5:
　Ⓒ + 5 = 11 in.

Ⓓ . . . 40 ÷ 8 = 5 in.

Add this to the given height of 6:
　Ⓓ + 6 = 11 in.

Therefore the areas of ⑦ and 49 are also the same height.

Area ⑦ is 49 in.²

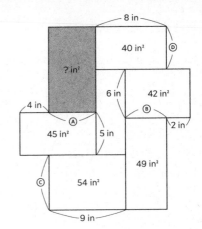

PUZZLE 287

Draw Ⓐ of equal height to the given area of 20.

Ⓑ must equal the given area of 18.

Ⓒ . . . 35 − 18 = 17 in.²

Ⓓ . . . (10 × 3) − 20 = 10 in.²

Ⓐ is double Ⓓ, so Ⓔ + Ⓑ must be double Ⓒ.

Ⓔ . . . (17 × 2) − 18 = 16 in.²

Draw Ⓕ equal to the given area of 34:
　Ⓖ = ⑦.

Ⓗ . . . Ⓔ + Ⓑ + Ⓒ − Ⓕ = 17 in.²

This is half of Ⓕ, so Ⓙ must be half of Ⓖ.

Therefore 18 must be triple Ⓙ: Ⓙ = 6 in.²

Ⓖ . . . 18 − 6 = 12 in.²

Area ⑦ is 12 in.²

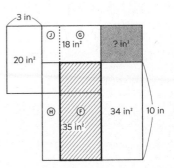

PUZZLE 288

Find Ⓐ + Ⓑ: 6 + 9 − 7 = 8 in.

This equals the given length of 8,
   so Ⓒ + 11 must equal 28.

Ⓒ . . . 28 − 11 = 17 in.²

Ⓓ . . . 34 − 17 = 17 in.²

This equals Ⓒ, so the area with stripes
   must equal 28.

Ⓔ . . . (28 + 28) ÷ 8 = 7 in.

Area ⑦ is (Ⓔ × 9) − 34 = 29 in.²

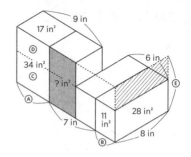

PUZZLE 289

Ⓐ . . . (7 + 6) − 11 = 2 in.

This equals the given height of 2,
   so area Ⓑ with stripes must be 44.

Ⓒ . . . (34 + 44) ÷ 6 = 13 in.

Ⓓ . . . 17 − 13 = 4 in.

This equals the given width of 4,
   and 34 is double 17.

Therefore 6 must be double Ⓔ: Ⓔ = 3 in.

Ⓕ . . . 7 − 3 = 4 in.

Find area Ⓖ with stripes:
   (17 − 2 − 3) × 4 − 22 = 26 in.²

From Ⓕ = 4, Ⓖ must equal ⑦.

Area ⑦ is 26 in.²

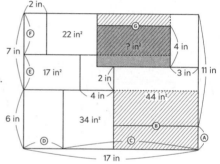

PUZZLE 290

Since 42 is double 21, and 10 is double 5:
   Ⓐ = Ⓑ.

Therefore Ⓒ = Ⓓ.

Draw Ⓔ of equal height to the given
   area of 23.

Ⓕ + 5 must equal 10, so Ⓕ = 5 in.

This equals the given height of 5, so the
   area with stripes must equal Ⓖ.

Ⓖ . . . 74 − 23 = 51 in.²

Length ⑦ is (Ⓖ − 21) ÷ 5 = 6 in.

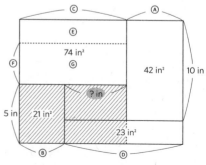

PUZZLE 291

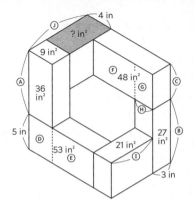

Ⓐ . . . 36 ÷ 4 = 9 in.
Ⓑ . . . 27 ÷ 3 = 9 in.
This equals Ⓐ, so Ⓒ must equal 5.
Ⓓ . . . 4 × 5 = 20 in.²
Ⓔ . . . 53 − 20 = 33 in.²
Ⓔ and Ⓕ are the same height and width:
   Ⓕ = 33 in.²
Ⓖ . . . 48 − 33 = 15 in.²
Ⓗ . . . 15 ÷ 5 = 3 in.
Ⓘ . . . 21 ÷ 3 = 7 in.
Ⓙ . . . 7 + 3 = 10 in.
Area ⑦ is (Ⓙ × 4) − 9 = 31 in.²

PUZZLE 292

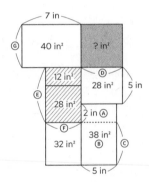

Ⓐ . . . 2 × 5 = 10 in.²
Ⓑ . . . 38 − 10 = 28 in.²
This equals the given area of 28 with height 5: Ⓒ = Ⓓ.
Ⓔ . . . 5 + 2 = 7 in.
Add the areas with stripes: 12 + 28 = 40 in.²
This equals the given area of 40 with width 7: Ⓕ = Ⓖ.
The given area of 32 has sides Ⓒ and Ⓕ.
These equal sides Ⓓ and Ⓖ of area ⑦.
Area ⑦ is 32 in.²

PUZZLE 293

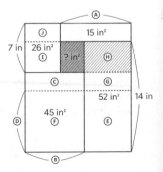

Since 52 is double 26, and 14 is double 7: Ⓐ = Ⓑ.
Draw Ⓒ of equal height to the given area of 15.
Ⓓ . . . 14 − 7 = 7 in.
This equals the given height of 7,
   so from Ⓐ = Ⓑ, Ⓔ = 26 in.²
Ⓕ . . . 45 − 15 = 30 in.²
This is double Ⓒ, so Ⓔ must be double Ⓖ: Ⓖ = 13 in.²
Ⓗ . . . 52 − 13 − 26 = 13 in.²
From Ⓐ = Ⓑ, Ⓗ = Ⓘ.
Ⓙ . . . 26 − 13 = 13 in.²
This equals Ⓘ, so the area with stripes must equal 15.
Area ⑦ is 15 − Ⓗ = 2 in.²

PUZZLE 294

Draw an area with stripes equal to 13: Ⓐ = Ⓑ.
Therefore Ⓒ must equal the given area of 15.
Ⓓ . . . 31 − 15 = 16 in.²
Ⓔ . . . (3 × 7) − 13 = 8 in.²
This is half of Ⓓ, so 3 must be half of Ⓕ: Ⓕ = 6 in.
Ⓖ . . . 36 ÷ 6 = 6 in.
This equals Ⓕ, so Ⓗ must equal Ⓒ.
Ⓗ . . . (7 × 6) − 31 = 11 in.²
Area ⑦ is 11 in.²

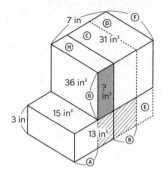

PUZZLE 295

Split the puzzle along the dotted line.
  From the given lengths, both parts
  are the same size.

Rotate the top 180°, then stack both parts:

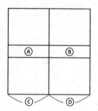

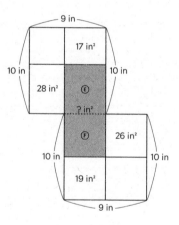

Ⓐ . . . 19 − 17 = 2 in.²
Ⓑ . . . 28 − 26 = 2 in.²
This equals Ⓐ, so Ⓒ must equal Ⓓ.
Therefore Ⓔ = 28, and Ⓕ = 26.
Area ⑦ is Ⓔ + Ⓕ = 54 in.²

PUZZLE 296

The given lengths of 7 are equal: Ⓐ = Ⓑ.

Two given areas of 36 have equal width: Ⓒ = Ⓓ.

Therefore ⑦ = Ⓔ.

On right, the given area of 36 is quadruple 9.

Therefore the area with stripes must be
quadruple 34: 136 in.²

Find Ⓔ + ⑦: 136 − 36 − 14 − 36 = 50 in.²

Area ⑦ is 50 ÷ 2 = 25 in.²

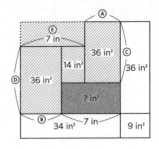

PUZZLE 297

Two given areas of 15 share a side: Ⓐ = Ⓑ.

Therefore the area with stripes must equal 50.

Find the area with crossed stripes:
(9 × 11) − 50 = 49 in.²

This equals the given area of 49: Ⓒ = Ⓓ.

From the given area of 15 with side Ⓓ,
Ⓔ must equal 15.

Therefore Ⓕ = Ⓐ.

Therefore ⑦ + 15 must equal the
given area of 50.

Area ⑦ is 50 − 15 = 35 in.²

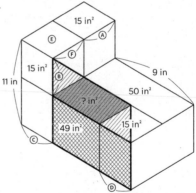

PUZZLE 298

Consider the puzzle in two parts, Ⓐ and Ⓑ.

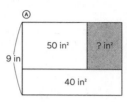
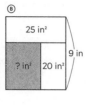

Rotate Ⓑ 180° and stack it on top of Ⓐ.

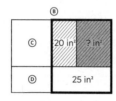

Ⓒ . . . 50 – 20 = 30 in.²

Ⓓ . . . 40 – 25 = 15 in.²

Ⓒ is double Ⓓ, so the area with stripes
must be double 25: 50 in.²

Area ⑦ is 50 – 20 = 30 in.²

PUZZLE 299

Consider the puzzle in two parts, Ⓐ and Ⓑ.

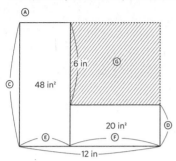

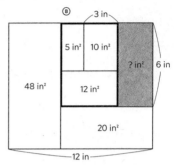

Imagine if Ⓒ grew longer:

Ⓒ – 6 = Ⓓ, so Ⓓ must grow.
Ⓒ × Ⓔ = 48, so Ⓔ must shrink.
Ⓔ + Ⓕ = 12, so Ⓕ must grow.
Ⓕ × Ⓓ = 20, so Ⓓ must shrink.
Ⓓ cannot both grow and shrink.
Therefore the overall proportions are fixed.

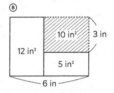

Flip and rotate Ⓑ to match Ⓐ.
By the same logic, the proportions are fixed.
Both given areas in Ⓐ are quadruple the matching areas in Ⓑ.
Therefore Ⓖ must be quadruple 10: 40 in.²
Area ⑦ is Ⓖ – 5 – 10 – 12 = 13 in.²

PUZZLE 300

Ⓐ × Ⓑ × Ⓓ × Ⓕ = 18 × 20 = 360.
Ⓐ × Ⓔ × Ⓒ × Ⓕ = 24 × 15 = 360.
Therefore Ⓑ × Ⓓ must equal Ⓔ × Ⓒ.
Ⓑ × Ⓓ = ⑦, and Ⓔ × Ⓒ = 16.
Area ⑦ is 16 in.²

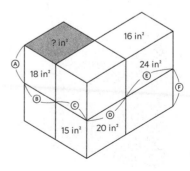